Rediscovering National Parks in the Spirit of John Muir

Rediscovering National Parks

in the Spirit of John Muir

Micha

The Unive

Sal

The Defiance House Man colophon is a registered trademark of the University of Utah Press. It is based on a four-foot-tall Ancient Puebloan pictograph (late PIII) near Glen Canyon, Utah.

19 18 17 16 15 1 2 3 4 5

Library of Congress Cataloging-in-Publication Data

CIP data on file with the Library of Congress

Printed and bound by Sheridan Books, Inc., Ann Arbor, Michigan.

Only by going alone in silence, without baggage, can one truly get into the heart of the wilderness.

—John Muir, in a letter to his wife, Louie

Contents

Preface and Acknowledgments

I recall in the fall of 1975 going to East Tennessee to help dedicate the Gee Creek Wilderness in the Cherokee National Forest. I found that it covered less than 2,500 acres but presented a beautiful fragment of Appalachian Mountain forest. At the dedication ceremony, a commemorative marker was unveiled, with the date, October 4, 1975, and these words from John Muir:

> Climb the mountains and get their good tidings. Nature's peace will flow into you as sunshine flows into trees. The winds will blow their freshness into you and the storms their energy, while cares will drop off like autumn leaves.

I think it's hard to read such lines in a setting to match and not be moved in an emotional and religious, or spiritual, way. After all, God marks the fall of a single sparrow and grieves the suffering of the humblest of his or her children. I myself am not religious—not in a structured, conventional way—but in a natural sanctuary like Gee Creek, or in Yellowstone or the Grand Canyon—or in any national park or forest—I feel I am in the presence of the sacred, and I embrace it. But, then, our paradise is fashioned on earth, the here and now, the one part of the infinite in which man and woman can demonstrate themselves as worthy.

John Muir felt uplifted and exalted in the wild sanctuary. Wild nature to him was an expression of God on earth—the mountains, God's temples; the forests, sacred groves. "But let children walk with Nature," wrote Muir, the father of our national parks. "Let them see the beautiful blendings and communions of death and life, their joyous inseparable unity, as taught in woods and meadows, plains and mountains and streams of our blessed star, and they will learn that death is stingless indeed, and as beautiful as life, and that the grave has no victory, for it never fights. All is divine harmony."[1]

Muir wrote those lines after sleeping in a cemetery in Georgia during the course of his legendary thousand-mile walk to the sea, from Kentucky to Key West, Florida. Perhaps we all ought to sleep in a cemetery and listen to the voices there, absorbing divine harmony as the basis for restoring harmony to human life and human society.

John Muir has walked with me, and his spirit has guided me in my life and career. In the preface to his book *Our National Parks*, published in 1901, Muir wrote: "I have done the best I could to show forth the beauty, grandeur, and all-embracing usefulness of our wild mountain forest reservations and parks, with a view to inciting the people to come and enjoy them, and get them into their hearts, that so at length their preservation and right use might be made sure."[2]

On one hand, you cannot have healthy parks surrounded by a sick society. But on the other hand, the parks can provide a model of health for society. I see that as an integral part of the mission of all of our parks, natural and historic, from a local community green to Yellowstone, covering two million acres.

Americans became acutely aware of their national parks in autumn 2013 when the Republican majority in the U.S. House of Representatives shut down the federal government. Doors were closed at American shrines like the Statue of Liberty, Lincoln Memorial, and Gettysburg battlefield, as well as natural treasures like Yellowstone, Yosemite, and the Grand Canyon. When the immediate issue before Congress was resolved, and park gates and doors were reopened, some of the very politicians responsible for the damage wanted to place blame for it on the National Park Service and its personnel.

One of them, Representative Randy Neugebauer of Texas, later apologized for scolding a park ranger over keeping the World War II Memorial in Washington closed during the shutdown. In a letter addressed to the director of the Park Service, Neugebauer wrote, "I'm sorry for the way I spoke, not only because I put the Ranger on duty in an uncomfortable position, but also because my remarks were not an accurate reflection of the regard I have for the Park Service."

Neugebauer wrote in his letter that he recognizes public service as a calling and that he respects all who respond to it. "Debate and dissent are crucial in our democracy, but it should be done in a respectful manner. While that was my intent, I realize that it wasn't the outcome."

National parks are treasures that belong to us all, regardless of political preference or affiliation. To consider the issue in perspective, turn the pages of history: Senator Robert La Follette of Wisconsin said of Republican Theodore Roosevelt, "His greatest work was actually beginning a world movement for staying terrestrial waste and saving for the human race the things which alone a great and peaceful and progressive and happy race can be founded." In 1903, Roosevelt went west from Washington, D.C., to North Dakota, then on to Yellowstone and California, where he spent three days in Yosemite with John Muir. The first night they lay in beds of fir boughs among the giant trunks of Sequoia trees, listening to the Rocky Mountain hermit thrush and the waterfalls tumbling down the sheer cliffs. "It was like lying in a great solemn cathedral," wrote Roosevelt. "It was vaster and more beautiful than any built by the hand of man."

In my time, I was privileged to enjoy a warm relationship with John P. Saylor, or "Big John," a senior Republican congressman from Johnstown, Pennsylvania. He was a towering man, standing six-foot-four, robust and yet wiry, a vigorous, powerful presence in any group. He was a progressive, but not in all things. He may have received the John Muir Medal from the Sierra Club, but he also received the distinguished service award from the ultraconservative Americans for Constitutional Action.

He was a creative and effective force at the center of nearly every major national park and conservation effort for two decades, including the Wilderness Act, Wild and Scenic Rivers Act, Land and Waters Conservation Fund Act, and the establishment of the North Cascades National Park in Washington State. My personal appraisal is that he was motivated by good old-fashioned American patriotism. He was the prime mover in Congress in the establishment of Piscataway Park, embracing seven miles of the Potomac River on the Maryland shore in order to preserve the vista from George Washington's mansion at Mount Vernon.

But my observations over the years have shown me that all is not well in our cherished national parks. Park people I know—maybe not the majority, but many—have labored long and hard and care deeply. They are tired of watching while parks are politically degraded into popcorn playgrounds. They feel frustrated and unfulfilled by the institutional bumbling of their high-level officials. They hunger for leadership committed to ethical and ecological principles, and to open communication.

I've watched change. The park system began in the West when areas like Yellowstone and Yosemite were remote. Now the system extends into major American cities—Boston, New York, Philadelphia, Washington, D.C., Cleveland-Akron, New Orleans, San Francisco, and Seattle. The great old parks, once at the edge of civilization, are now easy to reach from wherever you may be. The scenery remains the same, but as the number of visitors increases, the atmosphere of wildness and sense of discovery are modified. Visitors are caught up in suburbia transposed.

With their many museums, "visitor centers," trailside exhibits, and guided walks, national parks constitute the single most important influence in reawakening our sensitivity to nature and history. They give a breadth of unity to the nation and furnish guidelines for protecting and restoring the environment beyond their boundaries.

Here we find a gallery of American treasures, an endowment of riches that makes the United States the envy of the world. You may not realize it, but the national park idea was born in America, and from here has spread around the world.

Today more than one hundred nations have their own systems of national parks. Much of the Galapagos Islands are protected as a national park. So are the great game ranges of East Africa and South Africa; the heart of the Southern Alps in New Zealand; the highland moors of Dartmouth and Exmoor, the Lake District and Peak District in England; the Vanoise high peaks in France and the neighboring Alps in Italy; and one-fifth of the little island of Bonaire in the Netherlands Antilles, appropriately named Washington National Park. All of these derive from an American idea, an American ideal.

I've learned that national parks cannot be considered all things to all people. Crowds, congestion, noise, intrusions of human-made structures, and pollution of air and water all interfere with the appreciation, understanding, and enjoyment of the natural scene. To be fully enjoyed, the parks must first be fully preserved.

The answer is not simply a matter of where, but of how—how to look at national parks as sanctuaries, how to absorb the wonders of nature and history they preserve, how to reorient one's own way of looking and thinking. A slower pace expands the dimensions of time. Visiting fewer parks and staying longer at each one leads to better appreciation and more enjoyment.

You read a lot these days about what is wrong with national parks. I'm sure I have contributed my share to the chorus of criticism about them. But the larger truth is that there is far more right about our national parks and every reason to think, feel, and act positively. As I see it, if each person who visits the parks does a little bit, in his or her own way, the sum total of energy will help significantly. That is where I like to begin.

Many people have helped me in the preparation of this book, as they have helped over the years in my work in the national parks and in conservation. These have included rangers, naturalists, biologists, park superintendents, and others in the field, as well as policy makers in Washington. They have also included "book people," who made it happen, specifically John Alley, my editor at the University of Utah Press; Pamela Martin Popiolkowski, copy editor on early drafts; Virginia Hoffman on the completed text; and Glen Walters, friend and nearby neighbor, who helped with assorted computer issues.

On park problems, politics, and complexities, I am grateful to Stewart Brandborg, former executive director, Wilderness Society; to Paul Pritchard, former president, National Parks and Conservation Association; and to Nathaniel P. Reed, former assistant secretary of the Interior and devoted champion of the Everglades. I am especially grateful to Dr. Alfred Runte, scholar and park historian, for our friendship and shared views over many years. I am indebted to the Association of National Park Service Retirees, specifically to Bill Wade and Rick Smith, and cheer them in their work. I will also cite here the work of Kurt Repanshek, founder and editor of the National Parks Traveler. Kurt is a skilled journalist who has creatively turned his talent to the National Park scene.

My gratitude also goes to the following: Vera Guise, Don Castleberry, Owen Hoffman, Riley McClelland, Alfred Runte, and Scott Silver. Finally to my wife, June Eastvold, for her patience and shared love of the parks and to my daughter, Michele.

I

Feet on the Ground, Eyes to the Sky

I should make clear early on that this is not what you would call a scholarly work. Nor is it objective. After all, my autobiography (published in 2007) is titled *Rebel on the Road: And Why I Was Never Neutral.* I want my work here to reflect my labor and love for national parks and pure bias for their preservation.

There is a difference between studying the parks and being there, in the parks, as an ongoing life experience. I was especially struck by this idea on a day in April 2009 when I spoke at the Spring Wildflower Pilgrimage in Gatlinburg, Tennessee, gateway to the Great Smoky Mountains National Park. It was a special occasion, in celebration of the seventy-fifth anniversary of the establishment of this beloved national park. I believe I was invited as the author of the book *Strangers in High Places: The Story of the Great Smoky Mountains* (1966) and as an advocate for preservation. This includes actively working for wilderness designation in the Great Smoky Mountains, with serious criticism of the agency proposal to skirt its responsibility.

I found myself on a panel with two young history professors, a man and a woman, from nearby universities. It seemed to me they had dates and data associated with the history of the park but lacked the feeling that comes from knowing the place with intimacy and emotion. And neither ever mentioned the word "wilderness" or the effort to protect it in fact and law.

In contrast, I chose to say in my remarks that, for me, coming to the Smokies was like coming home, even though I was not a mountain boy, or a country boy, but a city boy. I felt that I could cite John

Muir's words: "The mountains are fountains, not only of rivers and fertile soil, but of men. Therefore, shall we feel that in some sense we are all mountaineers and going to the mountains is going home?"[1]

Hiking, being there, and standing up for protection and preservation breeds enthusiasm, idealism, love of life. As Benton MacKaye wrote in "The Appalachian Trail: A Guide to the Study of Nature," an article in *Scientific Monthly* of April 1932: "Primeval influence is the opposite of machine influence. It is the antidote for overrapid mechanization. It is getting feet on the ground with eyes toward the sky—not eyes on the ground with feet on a lever. It is feeling what you touch and seeing what you look at."

We are fortunate in America that we have such places, even with their various human-inflicted wounds and blemishes, at this advanced stage of history. National parks, national forests, national wildlife refuges, state parks and forests and county and city parks are all treasures to cherish and safeguard. The personnel who work at these places work for the long-range good of the country, and we are grateful to them. As Paul said in I Thessalonians: "We always give thanks to God for all of you and mention you in our prayers."

But, as I said at the wildflower pilgrimage, and elsewhere, no tract of public land has its future assured simply with a label, nor because it has a staff of paid professionals in charge. Those before us have set an example of faith and hope, and of principle to guide the way ahead. Let it be recorded that our generation has cared and given to sustain a treasured heritage.

2

Institutions Breed Conformity and Compliance

In April 1988, I delivered a lecture at what was called the Distinguished Visitors Seminar at the University of Washington College of Forest Resources. I spoke to graduate students, about twenty-five of them, studying for doctorates or master's degrees in natural resource policy. They were all mature professionals, normally working for federal and state agencies, and I enjoyed dialogue with them.

I was invited by the director of the seminar program, Russell E. Dickenson, recently retired director of the National Park Service. I had known Dickenson for many years while he worked his way up through the Park Service ranks. At one point he was regional director of the National Capital Region, dealing with the White House, Congress, and assorted powerful interests with their requests and demands, based on both their political needs and public service. He was deputy director under Ron Walker, who had been chosen by President Richard Nixon as Park Service director without any qualification. Dickenson was expected to hold Walker's hand and guide him along the way, but clearly he also waited and watched while Nixon fell and Walker went, too. Dickenson then became Pacific Northwest regional director, headquartered in Seattle, before returning as director in 1981. I knew Dickenson as a park professional who meant well and did his best against the odds presented by political higher-ups in the Reagan administration who knew little about parks and cared less. He retired in 1985 and moved with his wife to Bellevue, a Seattle suburb. He died in 2008 at age eighty-four.

In my lecture I endeavored to discuss principles of public service as I saw them.

> At the outset, I would like to pay tribute to an old colleague of Russ Dickenson in the National Park Service, Herbert Evison, a native of western Washington (though he now lives in the East in retirement). Herb is ninety-five or ninety-six. As a youngster he volunteered to serve as a cook when the legendary Stephen T. Mather made his four-day pack trip through the northern part of Mount Rainier National Park, in 1919. While on that trip Mather sparked the idea of a Washington State save-the-trees organization, presently to be known as the Natural Park Association, with young Evison as secretary. It was a go-getting group that convinced the legislature in 1921 to establish a state park system. Subsequently Herbert Evison became executive secretary of the National Conference on State Parks, and, later, for many years, an effective and outstanding chief of information of the National Park Service in Washington.

Following my dismissal in late 1974 as conservation editor of *Field & Stream*, a periodical then owned by the Columbia Broadcasting System (CBS), I received a perceptive letter from my friend Evison. He wrote as follows:

> One thing is a cinch: In this day and age when every company is owned by some other company, it is virtually impossible to say ANYTHING of importance without stepping on somebody's toes and irritating the financial nerve, directly or indirectly. Who in hell would suppose that anything said in *Field & Stream* would arouse the instincts of CBS? In conglomerates, each part is bound—is virtually obligated—to scratch the back of all the other parts; I don't know of any single element of the American scene that can more insidiously affect our liberties and our freedom of speech.

It impresses me, in retrospect, that if my friend Herb had left out the sentence about *Field & Stream* and CBS, his statement would have been just as valid. Change the word "company" to "institution" and it's even better, for all organizations—whether private or public, profit-making or eleemosynary, academic, professional, government, or what have you—once they become large and self-perpetuating, inhibit, if not repress, individualism, self-expression, and imagination.

Institutions generally, by their nature, breed conformity and compliance. The larger and older the institution, the less vision it expresses or tolerates. This is manifestly evident in the case of wilderness. The

Wilderness Act of 1964 and subsequent legislation assigned four federal agencies—the Bureau of Land Management, Fish and Wildlife Service, Forest Service, and National Park Service—the specific mandate to protect, perpetuate, and champion wilderness. I can't think of a single one of them that has come close to meeting its obligation and opportunity, except, now and then, with rhetoric and reports.

I also said in my lecture, "Concern for ecology may be expressed as a principle, but scarcely as something practical in critical need of defense. The best defense, at least in my view, is an alert and alarmed public. But national parks personnel are generally inward-oriented and poor communicators. They know the public as visitor numbers, but not as decision makers. Woe unto the parks person who goes to the public with faith or trust in his or her heart. The parks person is a professional, which is how she or he learned to appreciate the values of ecology in theory, but conformity and compromise in practice."

I told my audience that a professor once said to me, "What this country needs are ideas bigger and better than money can buy." Yes, I said, professions should be the standard bearers of ideas bigger and better than money can buy. So I think they were, they must have been, in their early days, all of them: law, medicine, religion, journalism, forestry, education. But time has mellowed them all. A professional degree is more of a license to employment than a charge to serve humanity without fear or favor.

Natural resource professionals ought to be in the lead of the revolution of values, but compassion and emotion are repressed in the training of these professionals. I told them of a talk I had recently made before an environmental conference in Alaska, after which I received a letter from one of the participants. "Not once in the ten years I spent studying forestry and land management while getting a PhD did anyone ever speak about ethics," she wrote. "I think that is criminal."

It doesn't have to be that way. "Imagination is more important than knowledge," said Albert Einstein. Psychologist Carl R. Rogers, in pleading for a new kind of human science, cited the fear among graduate students in behavioral sciences of what he calls creative subjective speculation: "They do not recognize that out of such fanciful thinking true science emerges."

For myself, I see imagination and a subjective value system as a force enabling people to rise above sheer facts, which may not be so factual after all. To say it another way, the power of human life is in emotion, in reverence and passion for the earth and its web of life. I didn't think this up—it's an ancient idea. But contemporary society is obsessed with facts and figures, and with modern machinery providing access to even more numbers. Alas, the analytical type of thinking of Western science has given us power over nature yet smothered us in ignorance about ourselves as part of it.

Twenty years ago I observed the controversy over the North Cascades, then administered by the U.S. Forest Service, and whether the region should be designated as a national park. North Cascades National Park was established by Congress in 1968, with transfer of national forest lands to the Park Service. The Forest Service is woefully ill-equipped to face the twenty-first-century needs and desires of the American people. One of my friends, a regional forester who became a high official in the agency, confessed to me that he had received "good training, but a poor education." That was in forestry, of course. The broader the training, the better the scholar. It's the narrowness, the commodity production focus, of forestry that engenders an antiwilderness ethos.

I don't see much better coming out of the National Park Service. The history of its administration of the North Cascades is written with sorry chapters the agency would just as soon forget. Management plans come full of internal contradictions, with lip service to ecosystem protection but without dealing substantively with crucial issues. For example, there are major threats to the integrity of the ecosystem just outside the park boundaries. Many choice landscape sections in the Mount Baker–Snoqualmie National Forest abutting the national park are slated to be cut within the next five to ten years, yet this is not even mentioned.

3

My Friends Harvey and Carsten, Superlative Park Advocates

Shortly after I moved to Bellingham, Washington, in 1987, Harvey Manning came to visit. He gave me a copy of his latest book, *Walking the Beach to Bellingham*, published the year before. It was a charming work, filled with wisdom, humor, and surprises. He showed me on the back cover an excerpt from a review I had written of it: "Harvey Manning not only has been through many battles but has trampled all over the state's backcountry. He knows and loves it all and writes about it in his own idiomatic manner."

Later we spoke together at an Earth Day program at the University of Washington and I always looked forward to the newsletter of the North Cascades Conservation Council, which he and his wife, Betty, edited.

Harvey was a writer, more or less like me, but maybe more so. After earning a degree in English literature from the University of Washington, he worked in public relations at the university and edited the alumni magazine. He loved the outdoors and went on to write immensely popular guidebooks for hikers and started Mountaineer Books. He was the Northwest hiker's hiker and scribe and might have acquired a degree of wealth. But material wealth was not his measure of success.

When Harvey saw the North Cascades slashed, burned, and roaded by timbermen and foresters, he and others began the campaign for a national park. He connected with David Brower of the Sierra Club, for whom no mountain was too high to climb, no battle of principle too tough to fight.

In 1957 Brower helped a handful of Washington State activists organize the North Cascades Conservation Council. "In the early 1960s, as it grew obvious we locals had to 'go national,'" Harvey later recorded, "Dave's leadership became paramount. He knew all buttons of all the players in the national game. He pushed them. And thus, in 1968 was created the North Cascades National Park."

Harvey Manning and Carsten Lien were two of a number of national park supporters, enthusiasts, citizen advocates, or what have you I met in and around Seattle and came to know as friends. When I connected with Carsten in the mid-1980s he was a visiting scholar at the University of Washington with an appointment to complete a thirty-year project of research and writing *Olympic Battleground: The Power Politics of Timber Preservation*, which proved to be a shocking, startling book, published in 1991. He knew the Olympics inside and out from experience and study, starting with his years working as a seasonal ranger-naturalist in the park in the 1940s. He once told me, "In the 1950s, I discovered the National Park Service allowed and supported logging in Olympic National Park; I knew there had to be a story behind it."

Carsten was a good friend, sturdy citizen, and champion of nature and the outdoors until his death in 2000 at age eighty-six. I knew him well over the years. He and his wife, Christi, came to our wedding at University Lutheran Church in Seattle on New Year's Eve 1993. More recently, while revisiting the Northwest, we went to see them at the retirement/convalescent home where they were living. In times past we had dinners with them and our mutual close friends, Alfred and Christine Runte. Or I would join Carsten and Al at a local cafe for a midmorning (or afternoon) coffee. I think it was a ritual for them to meet every day to analyze the state of the world, the state of Washington, and any other state that came to mind.

Harvey and Carsten were special people. So were others who made park advancement and enhancement their mission. I must say that everything about Harvey was idiomatic. In appearance, he looked like Grandfather Santa Claus but could not have cared less. Tracy Spring knew him well through her father and uncle, Bob and Ira Spring, famous photographers of Northwest mountains. She wrote me soon after he died: "Yes, the mold broke after Harvey Manning. If there's a heaven, Harvey and Ira, his old partner, continue their friendship and their debates, on endless hikes in good weather through pristine mountain wilderness."

Harvey Manning left a legacy and a challenge to citizens who enjoy nature the way God made it and who care about America the Beautiful and about themselves. He tells us that we need to express our care and concern with principle and vigilance. We need to stand up to be heard—loud and clear and strong enough to rally public support and to shape it into effective public policy and practice. This is not exactly easy.

Harvey passed on to his reward in Seattle on November 12, 2006. He was eighty-one, my vintage, and a friend—a comrade-in-arms if not quite a bosom buddy. I respected and admired Harvey's creativity as a writer and courage as an activist. He held principle high and stuck to it. On November 14, the *Seattle Times* gave him a cheery sendoff, calling him "a dedicated and caring conservationist," adding:

> The North Cascades National Park, the Alpine Lakes Wilderness Area and Cougar Mountain Regional Park all owe their existence to him. . . .
>
> The Issaquah Alps were nameless foothills south of Interstate 90 before Mr. Manning christened them in 1976. He was one of the original advocates of the Mountains-to-Sound Greenway, now a publicly owned 100-mile corridor of woods along I-90.

That is the least of it. Harvey's legacy really is in the standard he set. As for Carsten, his book on the Olympics stirred rave reviews but, as I recall, stony silence from the National Park Service. When I wrote to Al Runte for the facts of the case, he replied:

> Your reference is correct. The book has never been sold in the park. In the Park Service's words, "It is too technical for the general public." . . . However, two years ago, at the Hoh visitor center, every young naturalist poured out of the backroom to meet Carsten, and to have him sign their reference copy of *Olympic Battleground*. I had asked a young ranger whether they sold the book. "No, but we have it on the shelf in our office and refer to it all the time." Then I revealed that the gentleman standing by the bookshelf was none other than Carsten himself. "You mean that is Carsten Lien—THE Carsten Lien?" the ranger asked. "Yes." Minutes later, five young rangers surrounded Carsten standing with his walker, virtually to worship

at his feet. He left the Hoh with tears in his eyes. "I guess people do read my book," he said. "Yes, and the right people," I replied. "You have made a difference where it counts." Best of all, when they produced the "reference copy" for his signature, it was covered with underlines and marginal notations. In fact, it was almost falling apart it had been read so much. His book had made a difference indeed.

4

A Superintendent's View of Park Priorities

In his youth, my friend Carsten Lien worked as a naturalist at Olympic National Park and then went on to a career in business and civic affairs. Throughout his life he remained an Olympic advocate and defender, even when it meant bucking the tide of park administrators who felt obliged to support, or propose, one project or another inimical with safeguarding values of the resource in their trust.

In the 1980s I had a taste of it myself. A park employee had quietly alerted me to the upcoming construction of a power line across a wilderness study area without public review. That did not sound right to my source in the ranks or to me. It was against the law, as written in the 1964 Wilderness Act, and was against the spirit of protection as embodied in the law. So I drove to park headquarters at Port Angeles to ask the park superintendent, Bob Chandler, about it. He seemed taken aback by my failure to understand, and so his reply was simple rhetoric about park priorities: "Do you want to keep the ski development at Hurricane Ridge from opening on time this winter?"

That was the way he viewed priorities. Opening the winter season was more important to him than studying or protecting the wilderness in the area leading to it. I've heard the same line again and again throughout the national park system from administrators who believe that use must come first, and the more of it the better. Laws are meant to be bent or broken, depending on the judgment of the park administrator.

True, they do not all feel that way. Roger Allin, superintendent at Olympic in the 1960s, later recounted for me a conversation he had with his immediate superior, regional director John Rutter: "I was determined to get a wilderness bill for the park through Congress and so I discussed it with John. He reviewed my proposal and then said, 'You can get it through if you leave out these two areas.' That wasn't the point. I wanted to define and protect wilderness, not simply get it through Congress."

This leads me to recall that I came to the Olympic Peninsula for the first time in 1960. It struck me from the outset as a marvel of the original America. In terms of geography, the "upper left-hand corner" of the West was still in semi-isolation, beyond the beaten track, bordered by the waters of the Pacific, the Strait of Juan de Fuca, and Puget Sound. The mountains and dense forests had resisted civilization longer than the rest of the Pacific Coast. While similar forests parallel the ocean from Alaska to California, at that time none was more luxuriant than those of the Olympic Peninsula. It preserved a geological story, with about sixty living glaciers in the high country. Six major glaciers—some blue, some white—were on 7,965-foot-high Mount Olympus alone. But, then, Olympus received two hundred inches of precipitation each year, most in the form of snow.

I saw that, below the glaciers, the alpine meadows on the northern side of the park, a mile above the sea, were rich in mountain wildflowers, including glacier and avalanche lilies, white buckwheat, columbine, asters, and Indian paintbrush.

I learned that growing conditions in the moderate climate have been so favorable since the last glacial age that the Olympic Peninsula below the meadows has bloomed into a forest of evergreen giants. The national park embraces record-making trees, or near-record-making trees, of the "big four" Olympic species: Western red cedar, Western hemlock, Sitka spruce, and Douglas fir. The last of these, the "Doug-fir," growing to heights of 250 feet or taller, is the mightiest of the four, exceeded only by the redwood of California.

Actually, on that 1960 visit I did not come to see the national park, but to observe forestry as it was practiced under jurisdiction of the U.S. Forest Service. That agency was established late in the nineteenth century in response to the theft of millions of acres of federal domain and ruthless overcutting and devastation of great forests in California and the Northwest. The worst of it was described in *Looters of the Public Domain*, a tell-all written by an insider, Steven A. D. Puter, and published in Portland in 1908.

At Camp Grisdale, near the town of Shelton, I saw the logging operation of the Simpson Timber Company on its own land and adjacent land administered by the Olympic National Forest. Heavy harvesting of timber had begun in the late nineteenth and early twentieth centuries. In those days, loggers used hand tools, and the operation took them hours or days. But now, everything was accelerated, with modern diesel-powered yarders and whining, gasoline-driven chainsaws taking but a fraction of the time.

Loggers from Camp Grisdale attacked some of America's toughest, steepest, wooded slopes that once were considered beyond reach. They cut down virgin stands of the most luxuriant forests of the Pacific slope, if not of the entire world. Only a few of the great stands were protected in the national park, while heavy logging continued on private lands and areas administered by the Forest Service.

While out there on the ground, I could not help but turn my thoughts to preservation in the national park. In my book *Whose Woods These Are*, I wrote:

> Nevertheless, the national park represents the last glorious opportunity to preserve an estate of giants—trees, elk and bear, and all wild things, free to follow their natural course with a minimum of manipulation or interference. Here on the Olympic Peninsula, Americans are proving they can afford to reserve 890,000 acres for the natural sciences and the national soul.[1]

The idea of a national park had its proponents, but it was a long time coming, in the face of an entrenched culture in western Washington committed to timber, first and foremost. Theodore Roosevelt exercised his presidential authority in 1906 to establish Mount Olympus National Monument, but unrestrained logging continued all around it. Later, the Emergency Conservation Committee, headed by Rosalie Edge, a New York socialite and pioneer environmentalist, played a significant role in stimulating public awareness and concern. I will also cite the crusading work in the Northwest of Irving Brant, who turned his talent in journalism to the preservation cause. In the monograph "The Olympic Forests for a National Park," published by the Emergency Conservation Committee in 1938, Brant wrote these inspiring words: "Let this land, which belongs to the American people, be placed beyond the despoiling axe and saw, beyond the hunter's rifle, and we shall have for our own enjoyment, and shall hand down to posterity, something better than an indestructible mountain surrounded by a wilderness of stumps." Public desire for preservation grew until Congress voted to establish the national park, and President Franklin D. Roosevelt signed the legislation in 1938.

However, chambers of commerce, timbermen, and their friendly foresters now argued that if trees were cut and roads built, visitors would better see and enjoy the park. Legal and illegal logging continued, much of it authorized, approved, sanctioned, driven, and defended by Fred Overly, park superintendent from 1951 to 1958. In his book *Olympic Battleground: The Power Politics of Timber Preservation*, Carsten Lien recounts the "chainsaw gardening by the NPS," as well as how Overly dressed in a class NPS uniform but was really a hard-core logger in disguise. By the time concerned citizens awakened the National Park Service and stirred it to remove Overly, he had directed the removal of approximately one hundred million board feet of timber from Olympic National Park.

I met Fred Overly soon after, when he was posted to his next assignment as superintendent of Great Smoky Mountains National Park. He was helpful in my research for my book on the Smokies and we exchanged views. I cannot forget his rhetorical question: "What's wrong with logging in a national park?" It reminded me of Bob Chandler demanding to know if I wanted to keep the ski development at Hurricane Ridge from opening on time.

In time, Overly left the Park Service and returned to the Northwest to work as regional director of the Bureau of Outdoor Recreation, another Interior Department agency. We met several times and I noted he had not changed, and was now calling for reducing the size of Olympic National Park.

Now, turn the page. In early November 1998, I was the main speaker at a banquet in Seattle commemorating the fiftieth anniversary of the citizen group Olympic Park Associates. I felt privileged and honored to speak to members of an organization that had fought so hard and so well for a cause so worthy. I said at that event:

> National parks, when they were new, yielded discovery, adventure, and challenge. They should always do that. As the rest of the country becomes developed, and supercivilized, national

> parks should be held apart, safeguarded to represent another side of America, free of technology, free of automobiles, snowmobiles, and flightseeing, free of commerce and crowds, free of instant gratification, reflecting a pioneer, self-reliant side of America.

In my mind's eye, while speaking these words, I saw valleys along the Pacific within the park—the Bogachiel, Hoh, Queets, and Quinault, botanical gardens called rain forests. I saw the rain forest enriched by towering trees, shrubs, flowering plants, giant ferns, mosses, lichens, fungi, and animals, with air plants growing from the limbs of trees, even from the tops of the highest trees, draped in fragile beauty like mysterious poetry of the forest.

And along the fifty-mile ocean strip, I saw species of sea lions, seals, and whales on the rocks or in offshore waters, with thousands of marine birds nesting on the islands. I visualized endless waves and seascapes in the islands, along with rocky arches and crescent beaches. I said that I hoped Olympic Park Associates and other citizen groups across the country would lead the way in bringing something new to bear in public conscience and in politics. I said the sheer power of entrenched, interlocking institutions may make the challenge seem impossible, and yet individuals working together, or sometimes even alone, have worked miracles, as plainly evident in the national park we had come to celebrate.

5

"Politics is the Problem, and Getting Worse"

Gary Gray was a prize student of mine at Western Washington University. He lost his life on August 19, 1991, in a mountain climbing accident on Mount Shuksan in nearby North Cascades National Park. His body, at first buried under an avalanche of ice and snow, was discovered three years later. His mother, Shirley Gray, though not a writer by training, felt inspired and challenged to record his story, and hers, in *Roaring Mountain*, a book published in 2012. In it, she wrote:

> I was impressed by the warmth and kindness of Bill Lester, the wilderness district ranger, who had the unpleasant job of explaining to us in detail the accident and the rescue attempts that followed. When Curt [Gary's climbing partner] and Gary failed to return from their climb and the rescue was initiated, Bill Lester and Ranger Kelly Bush were the ones who had flown in the helicopter to search for them and found Curt. Bill described the frozen terrain where Gary was buried and showed us slide pictures of the accident site. His task was not an enviable one and I'll never forget his gentle, compassionate hug as we said goodbye that day. After we left Washington [State], he kept in touch, informing us of any attempts at recovering the body.

I was well acquainted with both Bill Lester and Kelly Bush, and was not surprised at Shirley Gray's account. I considered Bill and Kelly the best of parks people, wholly committed to their work in protecting the resource in their charge and in serving the public that comes to use and enjoy it. But I would not class them as typical of personnel at all levels throughout the parks.

In this same period I had ongoing correspondence with George Fry, who was then retired but had been superintendent of Great Smoky Mountains National Park in the 1960s, during the controversy over the proposed transmountain road. We exchanged views on park issues and policies. He wanted me to see him as a good guy, a partner in the cause of conservation, and not as a bureaucrat who had sold out. On August 28, 1992, I sent him the September 1967 issue of *Appalachian Trailway News*, in which I wrote a little essay, "A New Age of Walking Americans," based on a speech to the annual Appalachian Trail Conference at High Hampton Inn in North Carolina on May 20, 1967. In it, I wrote:

> Instead of devising new techniques to introduce nature to people and people to nature, they tend to follow the easy way, pouring concrete for more and wider highways, congesting visitors on treeless, barren camping suburbias—the kind of facility that could as well be located on private land—and utterly isolating them from the environment. In the process they destroy the natural features most people have come to enjoy. . . .
>
> The present park superintendent [George Fry] promotes this course of destruction. "None of us is getting younger," to quote his statement in the *Knoxville News-Sentinel* of November 2, 1966. "In fact, we're all getting older. So we can't all hike, visit the waterfalls or camp out. We must open new roads for visitors who are growing older. Many can't hike anymore."

Many years later, in continuing correspondence, Fry assured me he wanted to be on the same side, the right side, with me. On August 17, 1992, he wrote:

> I have just reread the Organic Act establishing the NPS, as well as Secretary of the Interior Franklin K. Lane's letter of May 3, 1918, to Director Mather. Down through the years these documents have been read, and reread, and often been quoted. The NPS, despite its problems and criticisms, has done very well. The basic problem, as I see it, has been politics, still is, and is getting worse. So you can't blame the directors and superintendents for all the problems. You can lay the problems at the feet of the Department of the Interior, the White House, the Congress, and the system.
>
> NPS needs to be an independent agency, free from all politics in Interior. The director should be appointed by the president

> for a six-year term, with the freedom to operate as are the directors of the FBI, CIA, and other agencies. He should be a professional type of manager, with proven natural resource management. . . .
>
> Even though top management posts in NPS are in the Senior Executive Service, only four of the eight regional directors are career managers. The speed with which William Penn Mott was dumped shows the trend of the Bush Administration. Now the latest I heard was that the latest appointee, the deputy regional director of the Pacific Northwest region was a state park–oriented individual, was a former Indiana state parks employee, no doubt a friend of [National Park Service director James] Ridenour. This is for the birds, no pun intended.

In some respects, Fry was right. The Park Service has done well in serving many millions of visitors. But it has not done nearly as well in protecting and preserving ecological resources in its trust. Likewise, George Fry may have scored well as a team player but not as a guardian of wild America.

In another letter, dated August 1, 1992, Fry told me that scouting, hiking, camping, and studying nature led to him earning a degree in forestry, and he learned to support the boss and be a team player. That was the way to get ahead and to survive. Fry wrote:

> Big George [National Park Service director George B. Hartzog Jr.] decided the Smokies would be an easy one to sell [as the first national park to be reviewed for wilderness designation under the 1964 Wilderness Act]. . . . That is when it hit the fan. All the purists, environmentalists and preservationists wrote 6,000 letters and publicly protested the road and putting the Smokies in wilderness. The park got a new master plan and there will be no more roads.

By the 1990s, Fry was well retired, but well informed on Park Service goings-on. The same holds true for everybody in the ranks, from top to bottom. They know the facts, the gossip, and the imperfections of their agency, the politics of parks. They learn to regard citizen conservationists as purists and avoid getting close to them. If they want to get ahead, or stay even, they must prove themselves as team players and steer clear of the purists.

Fry was correct about the fate of national parks under the Bush administration. Virtually every administration cleans house and distributes political plums among its own crowd. Some political appointees are qualified, but not necessarily. James Ridenour of Indiana was a patronage appointee of Vice President Dan Quayle. He brought with him Jim Walters, as deputy regional director of the Pacific Northwest. I met both and discussed park issues with them. They were trained in recreation, not in preservation. Ridenour had not actually been to the great national parks. When I met and talked with Walters in Seattle, it was clear that he saw little if any difference between national parks and state parks.

In his letter to me, Fry expressed a viewpoint widely shared in the ranks. He was absolutely correct in calling the basic problem political and getting worse. But to say that out loud while on the payroll—neither Fry nor anyone else would dare.

6

I Discover National Parks, the Parks Discover Me

My interest and activity in national parks began in the mountains of the southern United States. I had served in the military during World War II, worked as a newspaper reporter following my discharge, then joined the public relations department of the American Automobile Association (AAA) at its headquarters one block west of the White House in Washington, D.C. The job paid fairly well and had assorted pluses and minuses. I became the travel expert, writing articles for the affiliated motor club periodicals about one tour destination and another and cultivating contacts with writers and editors who might write and publish kind words about the association. The post-war period was a boom time, with new cars, new highways, new motels and hotels, new airplanes, and millions of new travelers exploring America and discovering national parks.

The Blue Ridge Parkway was new, too. My files show that I wrote my first article about the parkway—or about any national park, for that matter—in the *New York Herald Tribune* in April 1952. At that time, I was thirty-two years old and had very little experience in the parks. I recorded that, with newly paved sections, a total of 320 miles of the parkway would shortly be completed, out of the ultimate total of 468 miles between Waynesboro, Virginia, and Soco Gap, North Carolina, the gateway to the Great Smoky Mountains.

Construction had actually begun during the administration of President Franklin D. Roosevelt as a combination make-work and land conservation project, with strong interest and support from Secretary of the Interior Harold L. Ickes in his role as federal public works administrator. Work on the project had started September 11, 1935, when it was first called the Appalachian Scenic Highway.

At the time I wrote that article I knew little about conservation. My interest was in telling readers of the travel pages about some special place worth visiting. And that was what I found here. I quoted the park superintendent, Sam P. Weems: "The parkway is the longest road in the country—and probably in the world—designed solely for recreational travel, free of commercial signposting and traffic." I wrote further that, together with the two national parks that it links—Great Smoky Mountains of North Carolina and Tennessee and Shenandoah of Virginia—the Blue Ridge Parkway comprised the last large natural frontier in the eastern United States. I described the views from scenic overlooks and the footpaths in primeval forests and on mountain crests and the vestiges of old human culture. "Besides its natural attractions," I wrote, "the Blue Ridge is a treasure house of mountain folklore, preserving isolated old cabin homes and mills that reflect the lives and habits of rugged highlands pioneers."

In due course, as I saw and learned more, it struck me that the abundant and diverse plants, flowers, and trees must be to the Smokies and Blue Ridge what granite domes are to Yosemite; geysers are to Yellowstone; and the wide, deep chasm is to the Grand Canyon. The Blue Ridge Parkway came across to me as a public treasure if ever there was one, more than a scenic roadway—free of billboards and assorted commercial blight—but a composition of unspoiled natural, historical, and recreational areas.

National parks, I learned, had been virtually closed during the war but now were being deluged by millions of visitors, although facilities and funding were at inadequate prewar levels. Although I was unaware of it at the time, Bernard DeVoto, the historian and essayist, was working on what proved a classic piece on this subject: "Let's Close the National Parks," published in *Harper's* in October 1953.

I met DeVoto several times, at one conference and another, finding him as vigorous in person as he was in print. He had a broad nose that looked to me as though it might have been flattened by a punch. In 1953 we were in a seminar together at the Mid-Century Conference on Resources for the Future in Washington, D.C., where he enlivened and enriched the discussion by dueling verbally with a Wyoming politician who evoked ideas on western public lands that later became the agenda of the Sagebrush Rebellion and the Tea Party movement. That conference accomplished little, but brought me in contact with a thousand people talking about how to value and measure the benefits from nature.

DeVoto was a model of somebody I wanted to be like. As a historian, he won the Pulitzer Prize in 1948 for his book *Across the Wide Missouri*. As an essayist, he contributed a provocative monthly column, "The Easy Chair," to *Harper's*, and made the most of it, critiquing, among other things, the activities of the FBI and book censors in Maine. He wrote for other magazines as well (including *True*, for which he wrote a celebrated article on how to mix a dry martini). With columns and articles bearing titles like "The Anxious West" and "The West Against Itself," he championed public lands that I knew little about.

In one of his pieces in *Harper's*, DeVoto explained the call to advocacy. Some judgments, he said, are quite simple: express yourself and you have nothing else to do. "But there are also judgments that require you to commit yourself, to stick your neck out," he wrote. "Expressing them in print obliges you to go on to advocacy. They get home to people's beliefs and feelings about important things, and that makes them inflammable." In the 1950s, DeVoto tackled a major issue of the time—the proposal by the politically powerful Bureau of Reclamation to build dams across the Green and Yampa Rivers in Dinosaur National Monument, a little-known wilderness in eastern Utah, his own native state. If Congress approved those dams, it might well set a precedent for intruding in other national parks as well.

In those days, few writers, or publishers of magazines or newspapers, felt such issues were important. Soon after Gaylord Nelson came to Washington in 1962 as a senator from Wisconsin, he delivered a speech warning of the need to conserve natural resources, which the *Washington Post* reported at the bottom of the obituary page like a dead issue. Soon after Richard Neuberger came to Washington in 1954 as a senator from Oregon, he made a comment criticizing the poisoning of squirrels on the White House lawn. The media en masse treated his remark as a very funny joke.

The first time I read about the Wilderness Act—or the Wilderness Bill at that time—was in the *New York Times* of May 15, 1956. In the column headed "Conservation" (on the back page of the Travel section), John B. Oakes reported that Senator Hubert Humphrey of

Minnesota was sponsoring legislation to establish a national wilderness preservation system. "The idea is certainly worth exploring," Oakes wrote, "if what is left of our country in a natural state is worth saving, as many of us believe it is." That stirred my conscience, and my curiosity to learn more.

Oakes told me how it worked at the *New York Times*. We became friends at some point following his appointment, in 1961, as editor of the editorial page of the *Times*, and would meet now and then, based on his personal, long-time interest in the environment. He told me of a conversation he had had with Turner Catledge, the managing editor of the *Times*. Oakes said, "Turner, the environment is getting to be big news. Why don't you assign somebody to cover it regularly?" But Catledge brushed him off. "When there's a story there, we'll cover it." In short, it wasn't important enough, and a good reporter could handle it easily as the need arose.

This may explain why the *Times* ridiculed *Silent Spring*, the landmark book by Rachel Carson, soon after it appeared in 1962. The *Times* called it "wholly inaccurate" and said it would "unnecessarily frighten the readers." But that newspaper was not alone. *Time* magazine asserted that accidental poisonings from pesticides were "very rare," though Carson and history have shown otherwise. That was the media at work: deriding Carson as a bird lover, fish lover, cat lover, and devotee of a mystical cult; reacting without serious research or investigative reporting; and more willing to heed the public relations material furnished by the pesticide industry than the substantive scholarly work of a diligent crusader.

The environment, as a subject to cover and as a life's work, was open and waiting for me, and I was moving, consciously or otherwise, to pick up on it. My budding interest in national parks fortuitously fit with my work at the AAA. For one thing, the parks were favorite destinations of auto club members. For another, the clubs had been involved many years before when national parks were first opened to auto travel. The AAA and its affiliated clubs came of age concurrent with the national parks and had a degree of collaboration with the National Park Service. As a case in point, the Automobile Club of Tennessee was instrumental in establishing Great Smoky Mountains National Park.

Through discussions with parks people, I learned that while parks were flooded with visitors, appropriations for them remained at inadequate prewar levels. The Park Service headquarters at the Department of the Interior was two or three blocks from the AAA and it was easy to make connections. I would meet and talk with Herbert Evison, an old-timer who was chief of information of the agency, and I carpooled (from my home in suburban Virginia) for a time with Ronald Lee, an associate director of the National Park Service, who lived nearby.

These conversations turned to what the AAA could do for the parks, leading to a dinner and dialogue hosted by the AAA on December 10, 1953, at the Metropolitan Club, an upscale men's club in Washington, attended by Douglas McKay, secretary of the Interior; Melville Bell Grosvenor, editor of *National Geographic* magazine; Laurance Rockefeller, the philanthropist in conservation and benefactor of national parks; two former senators, Gerald P. Nye of South Dakota and Burton K. Wheeler of Montana; National Park Service director Conrad L. Wirth and associate director Ronald Lee; and AAA officials and personnel, including myself.

At the dinner, Wirth distributed a brochure prepared for the occasion: "The National Park System: Present and Future." He used it as the basis of discussion on how to win friends and influence public policy. To summarize: during the years of World War II, the number of park visitors had dropped to about five million from about twenty-one million in 1941, the last prewar year. The regular appropriations were cut 75 percent while the special funds—such as Civilian Conservation Corps (CCC) and Works Progress Administration (WPA)—were gone completely. In 1946, total travel to the parks was up again to twenty-one million and continued to rise rapidly from then on. But the government paid very little attention, being preoccupied with the Cold War and military police actions. In every respect, the parks were deteriorating, while facing a fantastic backlog of repairs. Nearly every campground was subject to excessive damage because of continuous overuse. Roads and trails were in ill-repair. Overnight accommodations were insufficient in number and, in many cases, inadequate in quality. Private landholdings within parks obstructed maintenance and sound administration, and the parks faced the threat of real estate subdivisions inside their boundaries. At least half the housing for park personnel was substandard, aged, and obsolete. The statement declared further:

> Growing crowds of visitors, numbering forty-six million in 1953, intensify the problem of protecting the parks. Many park features are irreplaceable and are subject to vandalism and theft. These must be continually protected if future generations are to enjoy them. Park museum collections are housed in structures that must be protected from fire. Fires may be caused by both lightning and careless acts of man, the risk from the latter increasing in proportion to the increase of visitors and lack of protection personnel. Management of the large wildlife population is an added risk.

Wirth made a case that it was time to do something dramatic before the parks fell apart at the seams in disrepair, and their unprotected features fell from the pedestal of magnificence into obscurity under the weight of visitor numbers. Subsequently, the AAA and the secretary of the Interior cosponsored, in 1954 and 1956, what we called American Pioneer Dinners. The first was held at a downtown hotel. The second, held at the Interior Department, was attended by sixty members of the U.S. Senate and House of Representatives and two Supreme Court justices. The state of South Dakota sent buffalo and elk meat for the main course. Ansel Adams, the celebrated photographer, sent prints of his work in Yosemite for everyone present. After dinner, guests went upstairs to the department auditorium to see a film, "Adventures in the National Parks," prepared especially for the occasion by the Walt Disney firm. And the National Park Service distributed the booklet "Our Heritage," the first official presentation of Wirth's brainchild, Mission 66, a ten-year national park system development plan.

In response to Mission 66, Congress did indeed boost appropriations as Wirth had hoped. Considerable funding went to purchase private holdings within park boundaries and restore them to nature. Important new park units were established, including portions of Cape Cod in Massachusetts, Padre Island in Texas, and Point Reyes in California. Mission 66 resulted in a lot of building and construction of roads, parking areas, utility systems, marinas, visitor centers, and training centers.

Looking back and expressing a viewpoint of my own, Mission 66 brought many benefits. It also showed that money alone cannot work miracles, for Mission 66 placed a higher priority on visitor comforts and facilities than on protection of natural and historic systems. That may reflect the American way, but it might have been otherwise, had Americans understood fully what they were buying.

Edward Abbey criticized Mission 66 for its unwholesome influence of "industrial tourism" in the parks. He was well equipped to speak, having worked as a ranger in several national parks, and authored *Desert Solitaire* (1968) and other books about his experience and philosophy. Too bad he missed the advent of "motor nature trails" in the Great Smoky Mountains, enabling visitors to say they have been there without getting out of the car.

7

Parkways, Monuments, and Memorials Are National Parks, Too

The more I learned about national parks the more I wanted to know and to explore them all. They became a large part of my life. Luckily, I could pursue my interest in the parks as part of my work as a writer and publicist for the AAA. I traveled around the country for the AAA, visiting parks whenever I could and, at the same time, endeavoring to raise my sights in viewing the world around me.

In Santa Fe, New Mexico, for instance, I visited the ancient cathedral and learned about Willa Cather's novel *Death Comes for the Archbishop* (1927), which led me to wonder what that was all about. I found that Cather had been a journalist on newspapers and an editor at *McClure's* during the muckraking days early in the twentieth century, doing the type of work that I dreamed of doing. She felt a love of the land and advocated what today would be called the counterculture of antimaterialism. This was clear in her description of Father Latour, the young pioneer priest (later archbishop) traveling on horseback through Navajo country with his Indian friend, Eusabio. Riding with Eusabio was like riding with the landscape made human; when they left a site—the tree or rock or sand dune that had sheltered them overnight—Eusabio buried the embers of the fire and the remnants of food, unpiled the stones he had piled together, and filled the holes he had scooped in the sand. It was the Indian's way to pass through a country without disturbing anything.

In those days, I lived south of Alexandria, Virginia, near Mount Vernon. I was moved to submit an essay, "My Journey to Washington," to

The Christian Science Monitor, whose editors published it on the features page on May 19, 1951. It began:

> The birds were awake first and I heard them before the alarm clock, sweeter music, too! Otherwise, it was a quiet, tranquil springtime morning in northern Virginia. Beyond the window blinds the great oaks reached skyward—they've been around this place a long time and I looked out wondering which was the oldest and which the tallest. . . .
>
> But I had to be getting on: with shaving, dressing, breakfast and an early departure, following the dreamy trail of nature and history. The broad Mount Vernon Memorial Boulevard was lined on both sides with green lawns and Virginia's loveliest trees and shrubs, red and white, pink and green, all of them alive as spring and as cool as the morning.

I described my drive along the Potomac River through Alexandria and into the nation's capital. The Washington Monument and Jefferson Memorial came into view, then around the Lincoln Memorial, where I saw "the big Mr. Lincoln sitting calm in the recess of his temple, still generating somehow the spark of kinship and understanding between all who come this way and the living Lincoln's hope and prayer of another time." The parkway along the Potomac, the memorial bridge, and the monuments to presidents—they were national parks, too, and part of my everyday living.

Over time I came to know various people in Washington—both in and out of government—who were connected with the parks. One of them, Devereux Butcher, worked for what was then called the National Parks Association, from 1941 to 1959, variously as executive secretary and editor of its journal. He was more of an outspoken, critical outsider than an insider. He insisted that, since the parks were set aside as nature sanctuaries to be held intact for all time, there must be no conflicting activity within them, such as mining, logging, dam building, airport construction, or grazing, nor nonconforming, crowd-attracting facilities such as golf courses, swimming pools, tramways, ski lifts, or tennis courts. Butcher was a prophet in the spirit of John Muir, who believed the parks contribute to our spiritual well-being, like literature, music, and art in their highest forms, and thus require unending vigilance to preserve them for that purpose. That made sense to me, and I picked up the spirit.

8

John Muir Comes into My Life

Then John Muir came into my life. I was in Louisville, Kentucky, early in the 1950s, visiting Eugene Stuart, manager of the Louisville Automobile Club. He knew of my interest in national parks and encouraged me. But, then, he was different than most auto club executives; most of them came across to me as businessmen who you might meet at a Chamber of Commerce luncheon or at Rotary, while he was a civic-minded independent.

One day, at his suggestion, Stuart and I drove to Mammoth Cave National Park, about ninety miles south of Louisville near Bowling Green in south-central Kentucky, the region called the southern sinkhole plain. En route he told me that for many years the cave was in private hands; it was an attraction of wide renown, where visitors toted whale-oil lamps and ladies were advised that "bloomers or Turkish dress constituted proper dress." It was touted as "the greatest cave that ever was" and thus was seriously considered when a federal commission went scouting in the early 1920s for one new national park in the East. Ultimately, in 1926, Congress selected the three finalists—Mammoth Cave, along with Shenandoah and the Great Smoky Mountains—to become new national parks. I thought Mammoth Cave was an acceptable choice, especially while walking in huge underground rooms with travertine or cave onyx, in thousands of intricate and artistic patterns, shaped over a period longer than I could imagine.

On one evening, my host and friend Stuart invited me to dinner at an upscale club with an elder statesman named Tom Wallace. Before going, Stuart told me that Wallace had long been editor of the *Louisville Times* until retiring in 1948. He said that Wallace had attained national prominence for his outspoken advocacy for conservation, specifically in defense of Cumberland Falls from a projected power dam. Wallace began his part in the issue in 1925 and was unrelenting, with more than one hundred editorials and public speeches all over Kentucky and beyond. He felt it was his duty, reasoning that "an editor cannot be valiant in print and a valet at heart."

It was an uphill battle against an entrenched political power structure that favored the dam but, ultimately, after five years of campaigning, Wallace and his supporters prevailed. In 1930 the site was converted into a state park—the prize of Kentucky, as it turned out—and Wallace was widely lauded for his principles and leadership.

At dinner, it was clear to me that Wallace's concern for that issue was part of a deeper commitment. He spoke of one campaign and another: against stream pollution and for the preservation of forests, game, and fish. When talk turned to national parks, he criticized plans for multiple power dams in western parks, declaring, "One purpose is to give employment to the needy blind—to engineers who cannot see why everything should not be destroyed." He deplored artificial "improvements" on nature generally, and said: "All such things are an outrage."[1] Specifically, he felt that the Mount Rushmore memorial in the Black Hills of South Dakota—with its giant carvings of the faces of George Washington, Thomas Jefferson, Abraham Lincoln, and Theodore Roosevelt—did not belong. He challenged me to think about it, and later, when I went to see Mount Rushmore for myself, I agreed wholly with him.

9

No Glaciers or Geysers in the Everglades, but . . .

I went to Florida in the early 1950s to see and report on Everglades National Park for the AAA. It was then still a new park, dedicated by President Harry S. Truman in December 1947, less than ten years before. But first I stopped in Miami to call on John Pennekamp, a long-time editor and columnist for the *Miami Herald*, highly regarded as a conservationist and park advocate. When the reefs off Key Largo remained unprotected, beyond park boundaries, he warned repeatedly that seashells, corals, sponges, sea horses, and other marine life were being hammered, chiseled, and even dynamited from the reefs to provide souvenirs. In this way, he and his paper were instrumental in setting aside a seventy-five-mile offshore sea park that would be named the John Pennekamp Coral Reef State Park, the most popular park in the Florida state park system.

Controversy seemed constant and continuous in the Everglades. Over the decades, so-called land reclamation, achieved by drainage and bulldozing, plus the introduction of canals, levees, and pumping stations, robbed the country of much of its natural beauty and variety. In earlier times the Everglades, at the southernmost corner of the North American mainland, ranged from keys, or islands, in Florida Bay to tangled mangrove swamps, sawgrass prairies, thick pinewoods, and tropical "hammocks" of broad-leaved trees and shrubs. The area teemed with life; natural conditions made it home to birds, reptiles, and mammals that cannot be seen elsewhere. More than a million wading birds nested here. In this uncharted country, alligators had been plentiful for centuries. Then, from 1800 until the early 1900s, millions were slaughtered each year. Some were killed for their valuable hides, others by adventurers who found thrill in the kill. The prevailing attitude was that money, time, and effort should be spent on mastering, controlling, and reducing nature in order to make way for growth and development.

Despite encroachment by farms and urban development, at the turn of the twentieth century the Everglades still covered about half of its historic range. It was still considered remote, beyond reach, an eternity of tall grass and shining water, an infinity of life-forms. The many varieties of birds of striking beauty became commercially valuable. Some, like the egret, were prized for plumes on ladies' hats, and so they were shot on a large scale and threatened with extinction. The National Committee of Audubon Societies raised funds to hire a warden, but three years later he was murdered. It was a tragedy that gave impetus to both bird conservation and preservation of the area.

Proposals to safeguard the Everglades were advanced early in the twentieth century while the conservationist Theodore Roosevelt was president. A small state park was established in 1916. Seven years later, in 1923, Stephen T. Mather, director of the National Park Service, declared that there should be an untouched example of the Everglades of Florida established as a national park, leading to a sequence of steps toward that goal.

When I called on Pennekamp in Miami, I was hoping to discuss the proposal to construct an overnight lodging facility at Flamingo in Everglades National Park. It was, I thought, a bad and harmful idea, disruptive of the ecosystem where the land meets the waters of Florida Bay; visitor needs would be better served by locating such a lodge on private land *outside* the park. Conrad L. Wirth, director of the National Park Service, had himself expressed much the same. Concessions in many western parks had come into being even before the parks themselves, but over the years they had generated serious headaches, conflicts, and controversies, which Wirth and colleagues hoped to avoid in new eastern parks, including the Everglades.

Pennekamp became angry with me for questioning the proposed construction. He showed me a large, loose-leaf notebook containing assorted resolutions and letters from local chambers of commerce, fishing clubs, and one or two airlines, all supporting the Flamingo project. I was taken aback, hardly expecting such a reaction, but endeavored to leave in good grace. Within the next day or two I was surprised to read a column by Pennekamp exposing and criticizing the influence exercised by an unnamed AAA spokesman over the director of the National Park Service and insisting the lodge must be built.

Pennekamp had his way. There was a lesson here for me. In 1959, the lodge complex opened thirty-eight miles inside the park's main entrance, with two stories, one hundred and three rooms, twenty-four cabins, and a campground, along with assorted headaches from the outset. Over the years, I visited and stayed there two or three times. In summer, the lodge was unbearably hot and humid, and air outside the screen doors was laden with massed mosquitoes in waiting. I was there once when Jack Stark was park superintendent (1971–1976). He felt so strongly the lodge was a mistake that he wrote me a letter spelling out chronic mismanagement and the addiction of employees to alcohol and drugs. Stark took a risk; a public official doesn't do that sort of thing. It didn't hurt his career, however, since he went on to be a regional director and superintendent of Grand Teton National Park.

Ultimately, in 2005, Hurricane Wilma—the most intense hurricane ever recorded in the Atlantic basin—ravaged Flamingo. The lodge and cabins were damaged beyond repair. Planning for a "new and improved" development began, aimed at making the complex smaller, greener, and more hurricane-resistant. But the idea is still questionable.

On that visit to the Everglades when the park was new, I called on the superintendent, Dan Beard, the very first in that position. Beard had been selected for the job in September 1947, and I felt he was a national parks person of the old school. He began his Park Service career in 1934, classed as a wildlife technician working on a Civilian Conservation Corps (CCC) project at Bear Mountain in New York.

I came to know him well, not only in the Everglades, but at Olympic National Park—where he was again the first superintendent—and in the Washington, D.C., office, where he was on the director's staff. I never heard him talk about it, but he was the son of another Dan Beard, a celebrity of an earlier day, well known as an illustrator, author, and social reformer, and even more known for his association with the Boy Scouts of America.

As for the Dan Beard I knew, when the signs were being planned for the lagoon at Royal Palm in the Everglades, the instincts of parks people were to tell the visitors not to fish there by posting signs reading

"No Fishing." Beard did not like the negative signs, so the signs finally read "Fishing is reserved for the birds." He was that kind of fellow.

During our time together, Beard referred to his earlier work in the Everglades in 1938 when he was classed as assistant wildlife technician. He mentioned a report he had written titled "Preliminary Thoughts on the Master Plan," and found a copy in the files to give to me. I read this excerpt:

> The southern Florida wilderness scenery is a study in halftones, not bright, bold strokes of a full brush as is the case of most other national parks. There are no knife-edged mountains protruding up into the sky. There are no valleys of any kind. No glaciers exist, no gaudy canyons, no geysers, no mighty trees, unless we except the few royal palms, not even a rockbound coast with the spray of ocean waves—none of the things we are used to seeing in our parks. Instead, there are lonely distances, intricate and monotonous waterways, birds, sky, and water. To put it crudely, there is nothing (and we include the bird rookeries) in the Everglades that will make Mr. Jonnie C. Public suck in his breath. This is not an indictment against the Everglades as a national park, because "breath sucking" is still not the thing we are striving for in preserving wilderness areas.[1]

Early parks people like Beard believed in preserving wilderness. That was their mission, what they strived for, and I cheered, believing the parks have been better for it.

10

Tough to Make a Living, Tougher to Say Something That Counts

When I left the AAA in 1957, I began writing for various newspapers and magazines, mostly travel articles about particular places, although increasingly, as time went on, about national parks. For the next thirty years of my career, I wrote for more magazines than I knew existed, from upscale to downscale, including *Parents*, *Parade*, *Smithsonian*, *Reader's Digest*, *Modern Bride*, *Travel Agent*, *National Parks*, and assorted airline magazines. On one morning I would finish and send off an article about national parks, and that very afternoon, the editor of another magazine would call to ask me for the same sort of piece—about national parks, of course.

I continually explored parks, monuments, battlefields, and historic sites, reporting on areas new to the public and new to me, as evident in my report from Chisos Basin, Texas, published in the *Chicago Tribune* on December 24, 1964:

> It's a notoriously long, long way from anywhere, across desert and scrubby grassland, into the bosom of the mighty Big Bend, and therein lies the beauty of a national park worthy of the name and worth all efforts to reach it.

At the approach of the fiftieth anniversary of the establishment of the National Park Service in 1916, an editor of Rand McNally called from Chicago and came to see me in northern Virginia. He explained that his firm wanted to publish a guidebook in commemoration of the anniversary and wanted me to write it. We talked awhile and I suggested we drive out to view and enjoy Shenandoah National Park, where he repeated the offer and we closed the deal. That guidebook was meant to be a "one shot" for the anniversary. We didn't get the first edition out until 1967, but a new edition was published annually for the next twenty-nine years, and I wrote every word in them. No, I did not get to all the national park areas, but I got to most of them, to some again and again. In the process, I met and listened to many different kinds of people: visitors from all over the country, visitors from abroad, students, seniors, rangers, naturalists, biologists, citizen park advocates, concessionaires, superintendents, seasonal employees, tourism promoters, and politicians.

My idea was to help readers become thoughtful, intelligent park visitors. In the opening pages of the Rand McNally guidebook, I wrote a section called "Planning a Park Trip, By Season and Reason." It began:

> National parks are more than places that can be identified by dots on a map. They are a way of travel. Visiting the national parks is truly an art, requiring time, training, patience. Walking through a gallery, the man who has learned how to look at pictures perceives deeper than eye level. He absorbs with his mind and senses. So too should it be with national parks.

During this same period, I was engaged by Kodak to write a guide to photography in the national parks. I was not much of a photographer, but they didn't ask me about that. I always thought reporting and writing constituted enough of a challenge and that I should leave photography to professional photographers. However, when a friend urged me to acquire and use a camera, if only to keep a visual record of the various places I'd been, I bought a pretty good one and accumulated lots of slides that I didn't know what to do with. I actually sold one photograph for a hundred dollars. Then, on a float trip through the Grand Canyon, I dropped my camera in the Colorado River. I recovered it, but it was damaged beyond repair. After that, I quit photography for good.

Midway through the Kodak project, I received a phone call from my supervisor complaining that my script was making it too difficult for people to take pictures. I didn't see it that way. I insisted that national parks must be treated by photographers with time and patience. I told him that I was writing a photography guide with an ecological approach, encouraging perception of native life communities—trees, birds, animals, fish, shrubs, flowers, soil, water, and air—all interdependent, dynamic, and changing. I counseled patience in photographing wildlife, considering that national parks are meant for animals on their own terms; unlike zoo animals, those in the wild move about with season, time of day, weather, availability of food, and other reasons they have chosen not to disclose. I shared my view and received no further complaints. That guidebook did well and Kodak came back to me for another one, about historic places.

I didn't go looking for trouble, but trouble and the misuse and abuse of parks kept presenting themselves and demanding attention. When I went to Gettysburg for the Sunday supplement *Parade*, J. Walter Coleman, superintendent of Gettysburg National Military Park and a historian by training, took me in hand. He wanted me to look beyond the usual battlefield story and patiently showed me the intrusions of development and commercialism on hallowed ground. This led to the cover story in *Parade* on December 14, 1958, titled "The New Battle of Gettysburg."

On the inside, the headline above my article read, "Neon signs, junk yards, dollar-grubbing are invading many great shrines. And the most hallowed of all is being desecrated in . . . the new battle of Gettysburg." In the article I wrote of how, on the three fateful days in July 1863, an estimated 160,000 Confederate and Union soldiers fought desperately for every inch in the greatest battle ever fought on American soil. The battle ended with staggering casualties: 50,000 killed, wounded, or captured. In autumn of that year, the nation mournfully dedicated Gettysburg National Cemetery, where 3,500 finally lay, and where President Abraham Lincoln delivered his immortal Gettysburg Address. And then I described the new invasion of commercialism that was obscuring the story of times past.

I followed that piece by writing an editorial for the *Saturday Evening Post*, published in September 1959. It was titled "Beer Halls at Gettysburg? Our National Shrines Need Attention!"

I wasn't through with Gettysburg. In "America the Beautiful—Heritage or Honkytonk?", published in the November 1962 *Changing Times*, I included Gettysburg in the list of assorted commercial gimmickry that tourists faced around the country:

> The new Howard Johnson motel advertises happily that it is "practically in the center of the Battlefield," inviting guests to experience the art of pleasant living only a few hundred yards

> from the field of Pickett's Charge and the scene of Lincoln's Gettysburg Address. Then there is America's most beautiful land of make-believe, a place called Fairyland, with monkeys and Mother Goose, "right in the center of things" facing General Meade's headquarters just below the National Cemetery.

In that *Changing Times* article, I also made a case about the road leading through the Black Hills to the Mount Rushmore National Memorial:

> The gimmicks are multifarious. The 35-mile road to Rushmore is an obstacle course of commercial attractions, all seeking to divert a million visitors, heading for a famous shrine, into their little crannies. The Reptile Gardens leads the way with the largest number of signs. There are four separate self-professed natural phenomena: Gravity Hill, Gravity Spot, Dizzyland USA, and the Cosmos (the Real Nature's Mystery Area).

Mount Rushmore itself was not much better:

> The mammoth souvenir shop carries the feeblest assortment of native handicrafts. The emphasis is on merchandising such items as nylon flags of all nations, Japanese-made plates bearing pictures of presidents and Mrs. Kennedy and figurines of Jesus Christ.

And then, in the same article, I wrote about Yellowstone:

> The sorry truth is that federal officials sometimes contribute to tourist blight, too. Despite the proud tradition and generally high standards of the National Park Service, the core of Yellowstone, the nation's oldest and largest, has deteriorated into a scandalous slum, the consequence of poor commercial management over a period of years and of soft and sloppy administration by park officials. . . .
>
> In short, 512-room Canyon Village, intended as a model in commercially developed lodging for the entire park system, has turned out to be a failure, faulty in design, faulty in construction, with poor heating, flimsy soundproofing and a multitude of other shortcomings. Rooms are horribly overpriced ($13.50 for two, $18.50 for four) and the entire setting is incongruous with the great park, landscaped largely with black asphalt and blinking lights over the cocktail lounge. The gift shop offers one of the worst assortments of trinkets in America, 8,000 separate

> items, principally cheap (but profitable) importations from the Far East, including imitation English Wedgewood, figurines in several colors, bells of Sarna and bongo drums.

Over the years, I wrote articles about the parks and their problems in the *New York Times*, the *Washington Post*, and the *Los Angeles Times*. "Parks in Peril" was the headline over my article in the *Washington Post* of April 12, 1981. It included this warning:

> Pollution of air and water, commercial encroachment, improper assignment and lack of field personnel to protect the resources are some of the critical problems that must be faced and solved in the 1980s—that is, if the parks are to remain unspoiled, peaceful havens for travelers to search for inspiration and wholesome recreation.

Over time I wrote several series in one periodical and another. On June 13, 1967, the first of five articles appeared in the *Christian Science Monitor*; the last, titled "Leisurely pace urged," appeared on July 17 that same year. It concluded:

> The visitor who has learned to explore thoroughly, to look at the treasures of the park deeper than eye level perceives and comprehends their meaning. He understands that the fundamental value of the parks is summed up in the one word: wilderness. He discovers in the wilderness a sense of freedom with personal fulfillment on the last frontier. He takes home with him enthusiasm, idealism, love of life, devotion to the parks concept.

A three-part series in the *Chicago Tribune* began on August 31, 1975. Titled "How long will the quality last?," it included this passage: "As millions of persons around the world seek out beauty, peace, and solitude on their vacations, the throngs of tourists as well as 20th century progress exact their tolls." The last article, published on September 7, 1975, was headlined "Proposals to save our parks." It included this paragraph:

> In one park, Shenandoah, in my state of Virginia, more than 700,000 pounds of garbage, including styrene plastic cups, paper plates, throwaway beverage cans, and sewage from the storage tanks of camping vehicles, were disposed of in a recent

> year at the landfill dumps inharmoniously placed in the wilderness. And where next when these dumps are filled?

No, I did not obtain this information from the superintendent or park headquarters, but they did not, and could not, deny it. I shudder to think of this practice widespread throughout the park system.

I wrote still another series, of five articles, under the heading "The Un-Greening of the National Parks," in *Travel Agent* magazine (published in the issues of August 18, August 21, August 25, August 28, and September 1, 1980). The editors introduced the series with this note above the first article:

> Travel agents have a big stake in national parks with concessioners and wholesalers turning to them more and more to book a public ever-hungry for some recreation, beauty and solitude. In this special series, Michael Frome, an eminent author and columnist, explores what's wrong with the parks and what must be done to halt and finally reverse a steady trend of deterioration.

That series earned the award for Best Magazine Article of the Year from the American Society of Journalists and Authors (ASJA), which I was invited to come to New York to receive. This I did on May 9, 1981, at a luncheon attended by several hundred writers, editors, and agents. When my turn came to accept the award, I said I was honored, and more:

> Freelance writers work alone, but they need each other too. They need to know that others are out there to share the struggles and successes with them. As soon as I qualified I joined the Society of Magazine Writers, later renamed the American Society of Journalists and Authors (ASJA) and, when I wrote my first book, the Authors Guild. The ASJA defines its role as offering "extensive benefits and services focusing on professional development . . . and, above all, the opportunity for members to explore professional issues and concerns with their peers." To my mind, that "above all" is above all.

I titled my remarks, "It's tough enough to make a living, but tougher to say something that counts." That is what I have tried to do with and for the national parks.

II

I Become an Author

One day in the late 1950s I was called by an editor at Doubleday in New York, which I believe at that time was the largest book publisher in the country. He said his company was considering publishing a domestic travel guide and, if I was interested in writing it, I should please prepare and submit an outline forthwith.

In due course, this led to the publication of my first book, *Better Vacations for Your Money: How to Get the Most Fun out of Your Vacation Dollar.* A Doubleday Book Club premium, it was an oversized paperback that sold for $1.95 when it came out in 1959. Another edition followed in 1960. I had no idea where it would lead—upward, as it turned out—but that was the beginning of my life as an author.

I asked Conrad L. Wirth, director of the National Park Service, to write the foreword, and he agreed. He wrote about the surge in travel by people going to see and enjoy the scenic beauty, the historic resources, and the rewarding recreational delights of our nation, plus some kind words about me.

Better Vacations proved acceptable and respectable, so Doubleday invited me to propose another book project. In retrospect, I may have been finding my way into conservation but was still thinking like a travel writer. Since Washington was a city I knew well and enjoyed, the result was *Washington: A Modern Guide to the Nation's Capital*, published in 1960. I explained my goal on the opening page:

> Six million or more visitors come this way yearly, yet there is so much they never get to see or realize about this wonderful

> town that belongs to all Americans, and really to all the world. For example, you can park a trailer two blocks from the Jefferson Memorial, watch a baseball game a little more than a home run's length from the White House, walk in wilderness in the heart of the city, board a submarine at harbor, hear superb chamber music at the Library of Congress.
>
> Washington, of course, is fundamentally a monumental city fashioned out of the dream of democracy. Here is where Lincoln walked and Washington before him; here were the local dramas of the Civil War and the War of 1812 (when the British burned many of the public buildings), the joyous inaugurals and the solemn funeral processions down Pennsylvania Avenue.
>
> Washington is everybody's city. The door is open to you at public buildings, museums, churches, and art galleries. Even the White House has its hours when the public is invited.

I recorded that there were then in Washington nearly 350 parks and monuments, including the White House grounds and the National Mall. I knew most of them. (At one time I had walked down from the top of the Washington Monument, passing fifty landings along the way and noting the carved tribute blocks within the shaft presented by states, cities, territories, colleges, churches, and foreign nations.) All told, those areas comprised 7,000 acres, or about one-tenth of the total capital land area. This included Rock Creek Park, nearly 2,000 acres in the heart of the city that Congress purchased for the nation in 1890—the same year as the establishment of Yosemite National Park—for its "pleasant valleys and deep ravines," as John Muir described it on a visit to Washington, "its primeval forests and open fields, its running waters, its rocks clothed with rich ferns and mosses, its repose and tranquility, its light and shade, its beautiful and extensive views."

I learned a lot while writing that book, and not solely about the subject at hand. I learned that parks are not a new, or even an American, creation. They were known in ancient civilization and in the development of European nations. In Europe many of the best of them were far from "the people's playgrounds"—the choicest were hunting reserves, privately held and only accessible to royalty and the rich. As Henry David Thoreau saw it:

> The kings of England formerly had their forests "to hold the king's game" for sport or food, sometimes destroying villages

> to create or extend them; and I think that they were impelled by a true instinct. Why should not we, who have renounced the king's authority, have our national preserves, where no villages need be destroyed, in which the bear and panther, and some even of the hunter race, may still exist, and not be "civilized off the face of the earth"—our forests, not to hold the king's game merely . . . but for inspiration and our own true recreation?"[1]

Thoreau was evoking a distinctly democratic American perspective. William Penn had the same idea with regard to the early settlement of the new world. He wanted Philadelphia to develop as a "greene country towne," and for every five acres cleared of forest he wanted one acre left wooded and undisturbed. As cities grew, open squares and commons were set aside for community purposes. City park systems emerged in the first half of the nineteenth century. In Philadelphia, extensive areas were acquired initially to safeguard the city's water supply and then Fairmount Park was officially founded in 1855, after 2,400 citizens petitioned for purchase of the Lemon Hill site. That was the beginning of the park system that now covers 9,200 acres, or ten percent, of the land in Philadelphia.

Early writers and artists sought not only to discover America but to define its spirit in a purely democratic American way. William Cullen Bryant, the poet, was considered bard of the early Republic, "father of American song." As a boy of seventeen in Massachusetts, he wrote "Thanatopsis," a poem that has helped generations of schoolchildren in their growing up, with these opening lines:

> *To him who in the love of Nature holds*
> *Communion with her visible form, she speaks*
> *A various language; for his gayer hours*
> *She has a voice of gladness, and a smile*
> *And eloquence of beauty, and she glides*
> *Into his darker musings, with a mild*
> *And healing sympathy, that steals away*
> *There sharpness ere he is aware.*[2]

As editor of the *New York Post* in his maturity, Bryant rejected the idea that parks were private, for the privileged alone. Thus, in the edition of July 3, 1844, he wrote an editorial titled "A New Park," in which he warned that booming Manhattan was in danger of losing its last chance to acquire territory for recreation and pleasure grounds.[3]

With the help of landscape architects Andrew Jackson Downing and Frederick Law Olmsted, Central Park was approved in 1851; it is now, by all odds, the most valuable undeveloped real estate in America.

George Washington, father of the country, shared Bryant's view and vision. Authorities and historians like to say that, in 1872, Yellowstone became the first national park. While that is clearly true, I learned that Washington, in his own way, contributed to recognition of natural beauty as part of our national heritage, which ultimately led to national parks. Washington was, by training and experience, a surveyor, farmer, and gardener, who valued time away from affairs of state at his home and plantation, Mount Vernon, overlooking the Potomac River.

The capital of the country, since the first Continental Congress and the Revolutionary War, had only temporary locations: in Philadelphia; New York; Baltimore; and Lancaster and York, Pennsylvania. But on July 16, 1790, Congress voted to appoint President Washington as a committee of one to choose a permanent location for a new capital and granted him authority to name three commissioners to lay out the federal capital district. Washington selected a favorite site a few miles north of his home. He then turned to Pierre Charles L'Enfant, who had served as a major on his staff during the Revolution, and appointed him the first United States city surveyor, or city planner. In an early report, L'Enfant wrote to Washington, "No nation had ever before the opportunity offered of deciding on the spot where the capital city should be fixed, or of combining every necessary consideration in the choice of situation."[4] He looked across the bogs and farmlands and envisioned an ageless city as beautiful as Versailles.

Washington supported the plan. Thomas Jefferson, as secretary of State, issued the order to L'Enfant to design the city, then met with him twice weekly. L'Enfant designed broad transverse streets and avenues, parks and circles, places for government buildings and even for the Washington Monument. It is considered a landmark in city planning. The government arrived in 1800, one year after the death of Washington, to a setting of parks, trees, and spacious outdoors. Two thousand varieties of trees—including thirty types of oak, flowering apple, quince, plum, and cherry—and shrubs grow in Washington today thanks to those early plans. No other city, and few arboretums, have such a wide range of flora.

As mentioned earlier, I lived near Mount Vernon and often visited there, so I knew it fairly well. I wrote that Washington was a progressive farmer who practiced crop rotation and once was awarded a silver cup for raising a prize donkey:

> Washington was an enthusiast of trees and gardens. His diary contains frequent references to the grading of the central lawn area and the transplanting of young trees and shrubs from the adjacent woods to the "Shrubberies" and "Wildernesses."

One might think that Mount Vernon, as a significant national shrine, would be maintained and protected by the National Park Service, but it's not. The Mount Vernon Ladies' Association was founded by Ann Pamela Cunningham of South Carolina, who noted while she rode by on a river steamer in 1853 that the place was going to seed. She first appealed to Congress to buy it, but without success. The money to purchase Mount Vernon from the Washington heirs was raised by public contributions and was acquired, in 1858, by the Ladies' Association.

Like Mount Vernon, some of the best historic places have been saved and maintained through the initiative of private citizens, organizations, and local jurisdictions. In my view, they are often well cared for. I remember in 1949, during my time in Washington, when the National Trust for Historic Preservation was established. David E. Finley Jr., director of the National Gallery of Art, was the first chairman, but Ronald F. Lee, an associate director of the National Park Service and a historian by training, was a prime mover. He told me he had been inspired by the highly regarded National Trust in the United Kingdom. Another historian, Fred Rath, left the Park Service to become the first executive director of the National Trust. From that beginning the trust has worked to "provide leadership, education, and advocacy to save America's diverse historic places and revitalize our communities." This includes the ownership of homes once owned by Presidents James Madison in Orange, Virginia, and Woodrow Wilson in Washington, D.C., plus almost thirty other properties designated as National Trust Historic Sites.

I visited the Woodrow Wilson House, the final home of the twenty-eighth president, on several occasions. It remains essentially as it was when he lived there and shows him as an educator, president, and world statesman. I was there one evening to hear an absorbing presentation on Wilson by Arthur Krock, who for many years was a celebrated *New York Times* columnist and editor. I believe that a recording of that talk, as well as other sound recordings and silent films, are still played for visitors.

I was there again May 25, 1971, my fifty-first birthday, to receive a Connie Award (for conservation) from colleagues and friends in the Society of American Travel Writers. They enlisted Representative Henry S. Reuss of Wisconsin, who was both a congressional leader and a friend of mine, to make the presentation. He read from the citation:

> Saluting the inspiration, distinguished leadership and significant accomplishments in the struggle against encroachment on precious preserves, natural resources, native wildlife and historic landmarks . . . and the tireless efforts to prevent despoilment of these resources and dilution of a quality experience for persons visiting and enjoying such properties.

I have tried, then and thereafter, to be that person.

12

Horace Albright

Portrait of a Conserver

Horace Albright's daughter Marian telephoned to say she had received the magazine, but that she was keeping it from her father as a surprise for the next day, his seventy-fifth birthday. That was fine with me, since I considered it a privilege to profile a hero and friend of mine who had played a particular role in the birth and growth of the National Park Service.

That article, titled "Horace Albright: Portrait of a Conserver," began: "Few living men, if indeed any, have exercised a longer, or more profound, influence over the destiny of treasured American lands and landscape than Horace Albright." It appeared in the May 1964 issue of *Westways*, a well-respected regional periodical. It was published by the Automobile Club of Southern California, but you could hardly tell the connection, since *Westways* featured some of California's best writers on subjects far removed from cranking up the car.

I first met Albright in Washington while I was still with the AAA. We stayed in contact and I went to visit him and his wife, Grace, in Los Angeles. I was not surprised to find that he continued with ideas and activities and with helping others to extend the long trail he blazed. He was generous to me in time and in inspiration.

The Albrights lived within a long stride of the UCLA campus, partially because one of Albright's current activities was membership on the Council of the Friends of the UCLA Library. Books were a passion to him. Filling the shelves, from floor to ceiling in almost every room of the Albright home, including the basement, was an amazing collection, carefully arranged to reflect the study of his special interests: the life and times of Theodore Roosevelt, mining, forestry, wildlife, the national parks, and the California story. These books, along with his correspondence and scrapbooks, comprised a prize package if ever there was one.

Albright helped me turn back the pages. In 1917, the Park Service itself was newly created only a year before. The entire staff of the bureau originally consisted of Stephen T. Mather,[1] Albright, and one stenographer. One year after graduation in 1912 from the University of California–Berkeley, Albright, the young attorney, went east to Washington in order to study mining and land laws on the scene where they are written. He joined the Department of the Interior at a time when the few national parks already established were administered loosely by the old General Land Office, with protection provided on the ground not by rangers but by the army. He was assigned to deal with legal matters concerning the parks; his duties shortly dovetailed with the movement led by the American Civic Association to provide these areas with a central administration.

In January 1915, Mather, age forty-seven, a man of accomplishment and influence, arrived as assistant to the secretary in charge of national parks. One of the first people he met was Albright, then twenty-four. Albright became Mather's assistant, protégé, and fast friend, and, when Mather was incapacitated with ill health, his stand-in.

Their first challenge was to draft the Organic Act to establish the new bureau to be known as the National Park Service. They became part of a group that met in Washington in 1915 and 1916 to review the draft legislation and plan strategy to get it passed. Participants included Frederick Law Olmsted Jr.; J. Horace McFarland; Representative William Kent of California; and Robert Sterling Yard, an old colleague of Mather's at the *New York Sun* (and best man at his wedding).

There were then eleven national parks, eighteen national monuments, and two other reservations, covering a total of 4.5 million acres. In that year, 1916, Sieur de Monts National Monument—forty-seven miles from Bangor, Maine, with surf-splashed cliffs rising to the highest point along the Atlantic Coast—was established as the first area in the East. It was later renamed Lafayette National Park and, in 1929, Acadia National Park. Within three years, Mount McKinley, Grand Canyon, and Zion National Parks were established.

The commitment and efforts of park advocates led to passage of the National Park Service Organic Act, or simply the Organic Act, of

August 25, 1916. It established the National Park Service, in a broad sense codifying the principles enunciated by Frederick Law Olmsted that Americans have the right to enjoy public scenery and the government has the responsibility to safeguard the scenery for them. The systematic protection of scenic resources marked a reversal of the prevalent view that the reserved public domain was meant for commodity production.[2]

Mather was determined to build a field force of professional rangers as a model of ethical and honorable federal employment. He refused to hire park superintendents based on patronage, insisting he would set the standards and pick his own people based strictly on merit. In Mather's day, operating funds were meager and promotions rare, but he vowed his agency would not be just another bureau. He inspired a spirit of mission and purpose, a willingness to stand against what he called "desecration of the people's playground for the benefit of a few individuals or corporations." Mather summed it up in a challenge for public support: "Is there not someplace in this great nation where lakes can be preserved in their natural state, where we and all generations to follow us can enjoy the beauty and charm of mountain waters in the midst of primeval forests?"

When Mather was stricken with an illness that would last a year, Albright was appointed acting director of the National Park Service. He faced the responsibility of finding office quarters, hiring personnel, handling the budget and complex legislation before Congress, and selling the principle of park protection to the public.

His background had prepared him for the challenge. He was born in 1890, the year the federal forest reserves were established, marking designation of portions of the public domain for protection in the public interest. His parents and grandparents were Nevada and California pioneers; as a boy in Inyo County, he rode the High Sierra trails with forest rangers, learning about camping and packing in the wild country. In due course he went north to study economics and mining law at the University of California as a member of the Class of 1912, in stimulating company, including Earl Warren, who became governor of California and then chief justice of the United States; Newton B. Drury, later a director of the National Park Service, as well as of the Save-the-Redwoods League; and Grace Noble, who was Mrs. Horace Albright until her death.

In those early years, Assistant Director Albright covered territory across the country. In 1919 he became the first civilian superintendent of Yellowstone, where the Army Corps of Engineers and cavalry units had been in charge since that park was created in 1872. He spent summers in the field, learning about the Yellowstone wildlife and backcountry, then he returned to Washington in the winter to pick up legal and administrative problems of the whole park system.

In 1924 he and John D. Rockefeller Jr. began their long period of association, founded on mutual respect and trust. The shy, wealthy gentleman from the East Coast arrived on a tour of Yellowstone with his three oldest sons, John, Nelson, and Laurance, to be greeted at the railroad depot by the park superintendent. At the moment of meeting, John III, eighteen, was recording in a notebook the amounts paid in tips to Pullman porters, while Nelson, sixteen, later vice president and governor of New York, was engaged in helping the porters transfer baggage from the train to Yellowstone park buses. Mr. Rockefeller was not unfamiliar with national parks, having earlier helped to acquire land at Acadia, in Maine, but this trip marked his full entry into philanthropy for conservation. Two years later he was back again and traveled with Albright from Yellowstone south to Jackson Hole, riding over an old wagon trail along the bluff overlooking the Snake River. It was an experience that resulted in his buying and donating land for Grand Teton National Park. On the same trip, he became interested in the redwoods of Northern California, providing $2 million to the Save-the-Redwoods League to purchase the famous Bull Creek Grove, later renamed by the California Park Commission as the Rockefeller Redwood Forest.

"I do not believe they could have been done at all without basic human compulsion to save what we prize for our fellows and for our children," wrote Albright of Rockefeller's accomplishments. "He brings to the problems of conservation a natural love of the beauties of nature, an alert, inquiring mind, a realistic sense of pace and appropriateness." The same could be said for Albright himself.

For thirty-five years, Rockefeller relied on Albright's advice and judgment in spending millions to safeguard precious lands for public use. Without this relationship, it is highly unlikely that certain national parks would exist today—not the Great Smoky Mountains National Park of the southern Appalachians, Grand Teton National Park of the

Wyoming Rockies, Virgin Islands National Park in the Caribbean, nor portions of other parks.[3]

In 1928 when Mather was taken critically ill and resigned, Albright was named director of the Park Service. In his four years as director, the Park Service turned its attention to preservation of historic sites, beginning with reconstruction of George Washington's birthplace at Wakefield, Virginia, then with the transfer of administration of military and battlefield parks from the War department. Though he knew how to get things done in the halls of Congress and in the executive branch, from the White House on down, he established a level of respect for his bureau based on pride and nonpartisan integrity. Politicians and promoters have always tried to obtain national park administration for their pet projects, regardless of their qualifications—it was far more difficult for them then than now.

A principal project of the early 1930s became the establishment of Great Smoky Mountains National Park, a half-million-acre Appalachian forest, much of it virgin wilderness, acquired for $10 million, one-half paid by the adjoining states of North Carolina and Tennessee, the other half by John D. Rockefeller Jr.

In 1933, after twenty years in government service, Albright resigned as Park Service director to become general manager, and later president, of the U.S. Potash Company. It seemed an unlikely switch. However, his family consisted of miners before him, and he had studied mining law at Berkeley. Besides, he probably wanted to get out from under Secretary of the Interior Harold L. Ickes. In any case, through the years of business, his influence in matters of conservation and parks did not wane. His headquarters in New York were only one building removed from those of Rockefeller in Rockefeller Center. He served the government on committees, commissions, and boards. Members of Congress counted on his wisdom and judgment. When I attended a testimonial dinner in Washington for Senator Clinton Anderson of New Mexico, a conservationist and champion of wilderness legislation, Anderson named Albright among a half-dozen or so who had been most helpful to him over the decades in Washington.

In his lifetime, Albright was associated with scores of good causes: in the East with the Hudson River Conservation Society, the Palisades Interstate Park Commission, the Theodore Roosevelt Association, and the American Scenic and Historic Preservation Society; and elsewhere with the American Pioneer Trails Association, the National Conference on State Parks, Jackson Hole Preserve, and the Museum of Navajo Ceremonial Art in New Mexico. In his own state of California he was involved with efforts to establish Kings Canyon National Park and to save Hetch Hetchy Valley in Yosemite, and he worked on behalf of the South Grove at Calaveras Big Trees State Park and with the Save-the-Redwoods League. Honors came to him from the government of Sweden; the Garden Club of America; the Sierra Club; his alma mater, the University of California—which named him Alumnus of the Year in 1952—and many others.

Yet it should be noted that he has not always been above criticism. In the early years of the national parks he supported public feeding of grizzly bears at Yellowstone and the "Rock of Ages singing" at Carlsbad Caverns, which verge more on entertainment than park conservation and are no longer practiced—though perhaps they may have been right in the context of their time. When asked if the Park Service had thought much about protecting ecological values, he responded, "We were busy trying to build a bureau and didn't know a thing about ecology." Once he said to me, "Now don't you pick on the concessionaires. The concessionaires are all right." So too with the peculiar placement and design of Jackson Lake Lodge in Grand Teton National Park, a misplaced obstacle in the path of one of America's noblest views; but this probably was Rockefeller's choice, based on his own ideas and the prerogative of his millions of dollars invested in protecting Jackson Hole.

I think of Albright's own words that the mission of conserving the best of America takes many forms of expression in different individuals, and that the real requirement is "wider support from more citizens who will take the trouble to inform themselves of new needs and weak spots in our conservation program." No man has done better to show the way to his own generation, or done more to inspire a succeeding generation. In his fearless, fighting days in Washington, a crusty old senator, Kenneth McKellar of Tennessee, once hailed Albright to Capitol Hill to demand construction of a road where wilderness belonged. When Director Albright refused to acquiesce, the senator exploded, vilified his parentage, and had to be restrained physically. That road, however, was not built. The qualities of courage and conviction, derived from his love of the land, have characterized the life of Horace Albright. Later in this memoir I will report on further encounters with him, but for now I say the nation will everlastingly be richer for his coming our way.

13

Drury in Defense of Dinosaur

I noted earlier the report on the Everglades National Park project prepared by Dan Beard in 1938. In the opening he quoted National Park Service director Arno B. Cammerer's challenge to see national parks not as they were but as they might be in a century of regeneration and regrowth. That was how parks people in those days viewed the parks and how they wanted Americans to view them as well.

In 1938 Cammerer delivered a speech in which he evoked a vision of wilderness restored in the Great Smoky Mountains. As he saw the future of fifty or one hundred years hence, the flowers, shrubs, and trees, even after the most destructive logging, would grow back as if little or nothing had happened. It would then be difficult, even for an ecologist, to ascertain whether a certain area was cut over, or that a certain mine had been worked years ago in a given location to the detriment of the park. In time such vision among administrators has become rare—their focus is on immediacy, political reality, and expediency—but in the Great Smokies and elsewhere, Cammerer's vision has been proven to work.

Cammerer, a Nebraska native, had joined the Park Service in its early years and had worked closely with Stephen T. Mather and Horace Albright. He was respected in Congress and committed to park projects in the East. In fact, he traveled over almost all the lands now included in Shenandoah, Great Smoky Mountains, Mammoth Cave, and Isle Royale National Parks, and worked hard to push the cause of the Everglades in Florida.

Cammerer's problem was that Secretary of the Interior Harold L. Ickes, the self-styled curmudgeon, did not care for him, and tried repeatedly to get rid of him. Ultimately he was spared the trouble. Cammerer suffered a heart attack and withdrew from Washington to be regional director in Richmond. He died a year later.

In those years one park professional followed another, which marks the difference between the original order and the new modern order. For Cammerer's successor, Ickes wanted a preservationist of high principle, regardless of the individual's politics. So, as the fourth director of the National Park Service, he appointed Newton B. Drury, a California Republican and classmate of Albright who had gained prominence as executive director of the Save-the-Redwoods League.

Drury was born in San Francisco to pioneer California parents. His father, Wells Drury, was a prominent western journalist. While attending Berkeley High School, Drury was a dance band cornetist, but once he entered college—the University of California, naturally—he switched to journalism, debating, and campus politics. In his senior year, Drury was elected president of the student body.

Following graduation, Drury remained for a time at the university as administrative assistant to the president and lecturer in English literature. However, he had a rendezvous to keep with his younger brother, Aubrey, a talented journalist and travel writer. Together they opened a San Francisco public relations firm, the Drury Company, and proceeded to assemble prestigious clients, like the Mark Hopkins Hotel, and likely could have spent their lives as successful and profitable image builders.

The redwoods interfered to change things. Taller than the first "skyscrapers" rising over the landscape of New York, the coastal redwoods (*Sequoia sempervirens*), in the early years of the century, still reached in glorious groves from the Oregon border to Big Sur. But they were in private hands and endangered as timber too valuable for their own good. Hoping to reverse the prevalent logging trend, prominent Californians and national figures of note created the Save-the-Redwoods League. When it came to getting things rolling, the league's founders turned to Newton Drury.

Drury was the man for the job; or perhaps I should say he and his brother were the men for the job. From the time the league was founded in 1919 until 1978, its history and the careers of the Drury brothers were virtually synonymous. As a matter of record, they would in time put together thirty magnificent redwood state parks comprising 135,000 acres.

Newton became the first executive secretary. He prepared the league's articles of incorporation and launched a public relations and fund-raising campaign on behalf of his clients, the trees. Over the years he conducted an eminently successful effort to enroll members and assemble donations for acquisition of virgin redwood forests. He became known for his guided tours of the Redwood Empire, on which he would induce citizens of means to purchase specific groves as living memorials. His tour for Mr. and Mrs. John D. Rockefeller Jr. and three of their sons, in 1926, helped lead to Rockefeller's gift of $2 million to buy the magnificent Bull Creek Grove in Humboldt Redwoods State Park.

Plainly here was a case of effective citizen action through an independent private organization. But it pointed out the need for a government role as well, if indeed the redwoods were to be saved and maintained for public good. Thus in 1928, Californians voted to support a state bond issue providing for matching funds derived from public contributions. This unique arrangement of public-private partnership marked the making of California's state park system. From its inception, Drury served as acquisition officer, putting the new parks together over the terms of four different governors.

He continued to fill his singular role in California until 1940, when he was invited to come to Washington as the fourth director of the National Park Service, following Mather, Albright, and Cammerer. Aubrey, meanwhile, took over as secretary of the Save-the-Redwoods League, so the Drury connection there remained unbroken.

The first time I met Newton Drury he impressed me as a man of poise and class, and that was definitely not on one of his better days. It was in the spring of 1951, at a luncheon in Washington the very day the newspapers reported his forced resignation as director of the National Park Service. He looked a little pained, as though smiling through the tears, but only a little. The larger expression was of pride in a tough battle lost, and of determination that the war was not over.

It wasn't easy. Director Drury had resisted plans to construct dams in national parks and had been forced out; but the dams were not built, and he successfully demonstrated that America's natural sanctuaries could be held inviolate, come what may.

Drury was guided by principle rather than by politics or expediency. "If we are going to succeed in preserving the greatness of the national parks, they must be held inviolate," he insisted. "They represent the last stands of primitive America. If we are going to whittle away at them we should recognize, at the very beginning, that all such whittlings are cumulative and that the end result will be mediocrity."[1]

Conrad L. Wirth, who later succeeded Drury as director, once spelled it out for me. He said that just as Drury assumed charge of the Park Service, war was in the wings. To make room for defense agencies, the Park Service and other bureaus were moved to Chicago, leaving only a skeleton liaison staff in Washington. Though for five years his headquarters was out of the main arena, the director faced a barrage of demands and pressures: repeated proposals called for sacrificing one or another park to the war effort for the installation of equipment on mountaintops, military maneuvers, or mining and logging. Wirth said, "I saw him in conferences when he would answer these proposals by saying, 'If you can bring a statement from the secretary of your department saying the war absolutely depends on exploiting a national park, we might give in.'" As a consequence, little damage was done to the parks, and none of it needlessly.

Following the war, new threats arose. Commercial interests brought political pressure to open Olympic National Park to mining and logging. These interests also demanded access to national monuments, including Saguaro, Organ Pipe Cactus, Death Valley, Joshua Tree, and Glacier Bay. The Bureau of Reclamation and the Army Corps of Engineers discovered potential in parks for water power, flood control, and irrigation and proposed projects in Big Bend, Glacier, Grand Canyon, Kings Canyon, and Mammoth Cave National Parks.[2]

Drury stuck to his guns in opposing two dams that the Bureau of Reclamation, a "sister agency" at Interior, wanted to build in Dinosaur National Monument. In June 1950 Secretary of the Interior Oscar L. Chapman (who succeeded Ickes) announced his support for the dams. The following spring he removed Drury. At that time, I knew little about the issue and did not understand it. In due course I observed how Dinosaur galvanized the environmental movement to defend it in a way that had failed to happen at Hetch Hetchy years before.[3]

During his long life (which ended at the age of eighty-nine in December 1978), Drury was many things, but a preservationist above all. Maybe it's the concepts he embodied that count most—the ideas that people can actually save big things seemingly beyond reach, then protect them without compromise or sacrifice. As he himself said, with typical clarity while fending off one threat after another to the national parks, "Civilization is encroaching on the wilderness all over the land. What remains of it becomes increasingly precious."

"He could have written extensively," his old friend Albright said at a commemorative dinner for the Save-the-Redwoods League's fiftieth anniversary on June 16, 1968, "but he dedicated his life to saving and protecting primitive, unspoiled features of America's heritage—not by words alone, but by deeds."

14

Hetch Hetchy Again, Now in Utah

Everywhere in America during the 1950s and 1960s, fresh blight spread across the landscape, and that was before the ascendance of McDonald's golden arches. Peter Blake, then editor of *Architectural Forum*, wrote a book, *God's Own Junkyard: The Planned Deterioration of America's Landscape* (1979), showing in all its kitsch the modern outdoor décor of billboards, neon signs, automobile junkyards, and uncontrolled ribbon development, degrading the landscape and public taste.

Tourist traps and roadside zoos commercialized nature, while phony Indian trading posts exploited indigenous culture. The Black Hills of South Dakota provided a classic case. The twenty-five-mile road from Rapid City to Mount Rushmore was an obstacle course lined with "family attractions." Animals caged at souvenir stands and gas stations, on animal "farms," and in animal "gardens" and animal "parks," virtually all of them displaying scraggly and mistreated deer, snakes, and bears. They were the poorest places to introduce children to wildlife.

In this same period the national parks were not above or beyond attack. As early as March 4, 1951, John B. Oakes, in his column "Conservation" in the *New York Times*, took up the proposed dams in Dinosaur National Monument in Utah, warning of the impending "stress on the promotion of organized recreation in our parks and monuments, and a slackening of interest in the preservation of untouched areas of wild and natural beauty for themselves alone."[1]

After Newton B. Drury's departure as National Park Service director, the fight over the proposed Echo Park dam heated up, emerging as the first major conservation issue of the postwar years. Dinosaur, in the remote reaches of northeastern Utah and western Colorado, had been designated a national monument in 1913 to preserve the skeletal remains embedded in the rocks, and then was enlarged in 1938 to more than two hundred thousand acres to preserve the bordering high cliffs and canyons above the Green and Yampa Rivers.

Western congressmen favored the project of their pet agency, the Bureau of Reclamation, despite its impact on the most famous dinosaur repository in the world. But Representative John P. Saylor, a Pennsylvania Republican, boldly spoke against it. It was his first big fight, propelling him into the front ranks of conservation. Environmental groups rallied against the Echo Park dam in a strong united voice. It was Hetch Hetchy again. The crux of the matter for them was not simply the integrity of Echo Park, but of all national parks, wildlife refuges, and other natural preserves. And for the first time national media focused broad public attention on a seemingly regional resource issue.

Bernard DeVoto wrote a well-documented article about the Dinosaur National Monument controversy in the *Saturday Evening Post* of July 22, 1950: "Shall We Let Them Ruin Our National Parks?" Subsequently reprinted in *Reader's Digest* in November 1951, DeVoto's article helped make Dinosaur a national issue. It encouraged opponents to intensify their campaign. It also led to a split, perhaps inevitable, between federal agencies and the citizen conservation movement.

Gilbert Stucker, a professional field paleontologist, made his first visit to Dinosaur in 1953 completely on his own. Normally he conducted studies for major museums (including the Smithsonian Institution and the American Museum of Natural History in New York) and universities, but, in the case of Dinosaur, he went primarily because of his concern over the proposed dams.

After reviewing the scene, Stucker returned east and called on Horace Albright in his New York office. Though long out of government, Albright remained connected and influential. In the course of their discussion, Stucker proposed using the dinosaur quarry as a positive project, a platform from which to generate public interest and concern.

"Not only would we start working back into the cliff," he explained, "we would expose them for public exhibit, showing people bones etched out in relief, bones of animals 140 million years old, in the rock where they died before the rock solidified. We would tell the public, 'You may think this is wonderful, but it's only a small part of the treasures in Dinosaur National Monument. Back in these canyons we have ocean beds with millions upon millions of sea life—shells and corals, coral reefs, all sorts of remains of prehistoric life.' We would get people interested in the quarry and, through it, interested in the canyons and the threat to them."

Albright sent Stucker to talk with National Park Service director Conrad L. Wirth. Wirth responded that he and his staff had been considering such a project. And the year following, when it began, Stucker was offered, and accepted, temporary appointment as ranger-naturalist. He would spend summers at the monument and winters working in paleontology at the American Museum of Natural History.

He convinced the National Park Service to build the unique visitor center displaying and interpreting tremendous bones of animals many millions of years old. Then, at the site, he would tell visitors about the ocean beds back in the canyons with millions upon millions of remains of sea life.

Stucker worked at Dinosaur as part of a four-man team—before public view on the north wall of the quarry—removing tons of rock with air-powered jack hammers and rotary rock drills, then roughing out bones with smaller chipping hammers and detailing them with hand tools. Visitors to the quarry would ask questions, giving him the chance to lecture to large and small groups. Some visitors asked surprising questions, like "May I pet a bone?"

Stucker and I became friends. Thanks to his encouragement, I visited Dinosaur and interviewed Tobe Wilkins. "There is no place like it in the world," Wilkins told me at the site. In 1953 he began working with Stucker on the rock wall and continued to do so for thirty-five years. (Stucker credits Wilkins with the greatest single contribution to the remarkable display of fossil remains.) "It's like unwrapping a Christmas present," Wilkins said. "You never know what you're going to find. You uncover a piece of fossil—and realize you are the first human to ever see it." Stucker and Wilkins typified a breed of committed personnel I met throughout the National Park Service.

"I realized that when I discussed the proposed dams, I was exceeding my authority," Stucker later recalled. "I was supposed to explain the quarry to the visiting public, not ask people to write their congressmen. At one point, Superintendent Jess Lombard called me into his office and said, 'I just had a telegram from the Secretary of the Interior directing that no Park Service employee is to discuss the threat of dams in Dinosaur National Monument. I know you've been talking against the dams. If you continue, I will have to separate you or discharge you.' But I knew I must talk against the dams and somehow rode it through until the question was settled and Dinosaur was saved."

Thanks to many individuals and organizations, Dinosaur indeed was saved. In 1956, after much debate, Congress eliminated Echo Park from the Colorado River Storage Project.[2]

Once it was over, Stucker was offered a permanent position with the National Park Service. He felt that sometime in the future he might want to speak out and would not be able to without risk. Instead he returned to the American Museum of Natural History, conducting field studies in various parts of the continent. On his own he continued his conservation work, joining efforts to add choice areas to the national park system, including Fossil Butte National Monument, in Wyoming, preserving fish forty or fifty million years old in laminated clays and silts.

Who can tell the full value of preserving these places? Paleontology is said to be a very careful, exact science, yet in years or generations hence, specimens preserved may reveal much more than they do today, because of advances in knowledge and human intelligence we cannot foresee. I pondered that question a few years after the Dinosaur issue was resolved, when two rangers took me to float the Green River into Echo Park. I saw and felt this large, deep, powerful river, from 100 to 300 feet wide, swirling and plunging, bordered by folded and tilted rock layers and deep canyons. The scene was haunting, wild, beautiful, and thoroughly natural. One of the rangers told me that archaeological evidence suggests these canyons and river valleys were once home to the ancient seminomadic Fremont Culture people, who left their rock art on canyon walls and in sheltered overhangs. Then the other ranger said that John Wesley Powell had rowed this way and named features such as the Gates of Lodore and Steamboat Rock. I felt that we must be floating through some of the most spectacular canyons in the United States, but the comment of that ranger reminded me that I had already been to Flaming Gorge in southwestern Wyoming. I believe that Powell named that one, too, but there is not much flame, or gorge either, since both now are submerged by the Flaming Gorge Dam and Reservoir.

15

Whose Woods These Are

In 1960, a fellow named Clint Davis, the director of information and education for the U.S. Forest Service, added a new dimension to my life. I had met him at various functions around Washington and had called on him at his office once or twice. I knew and understood little about the Forest Service or the national forests. I did know that the Forest Service was headquartered in the Department of Agriculture, and that the national forests were much larger and covered more of the country than the national parks, with substantial opportunities for recreation. I knew, further, that the national forests were open for commercial uses such as logging and grazing, but that they covered large areas of wilderness as well.

I knew this, at best, in a vague way. Davis had seen or read some of my published writing and apparently wanted to interest me in writing about the national forests. We talked a few times and then he invited me to join him on a horse pack trip into the Bridger Wilderness of Wyoming, south of Yellowstone and the Grand Tetons, with a group sponsored by the American Forestry Association called Trail Riders of the Wilderness. In addition, we would visit other national forests in the area and be gone for three weeks.

The trip turned out to be a magic time for me. The fifteen or so trail riders were city people, mostly professionals wanting a different kind of vacation. We rode uphill, downhill, alongside sheer rocky crests, and across streams and green meadows in the Wind River Mountains, one of the most rugged sections of the Rocky Mountain chain. We camped in woods and valleys abloom with wildflowers. At one point we came to the confluence of plunging streams forming the headwaters of the Green River. At 11,000 feet we were in the country explored by General John C. Fremont in 1842. He wrote later that here he was overcome by the power of natural stillness; and so was I. For Fremont it was "a concourse of lakes and rushing waters, mountains of rock, dells and ravines of the most exquisite beauty, all kept green and fresh by the great moisture in the air and sown with brilliant flowers." For me, now above timberline, beyond civilization and free of its artificial noises, close enough to touch stars, I felt connected with history and beyond history with eternity and infinity. It made little difference to me that it was in a national forest rather than a national park. It was America at its best.

While on the trip Davis and I talked at length about the national forests, the Forest Service, and the issue of preserving wilderness, which Congress was then considering. He didn't try to convince me of anything, as far as I could tell, but wanted me to see things for myself. I felt privy to something important here that might make a book. National forests were spread across the country; they had been established around the turn of the twentieth century to protect the public interest from exploitation by logging, grazing, mining, and other commerce. Since I knew little about them, how much could the public know? Besides, I was experiencing in the Bridger Wilderness a marvelous fragment of America that I hadn't realized existed, and there were more of them within this same system. I didn't know where or how many, but I believed them to be rich in history and adventure as well as scenery.

On my return home I wrote to Doubleday, reporting on my experience and outlining a book idea that became *Whose Woods These Are: The Story of the National Forests*. I was fortunate in that Davis found a provision in law directing the Forest Service to encourage public understanding of forestry. That became the basis of the help—the support, I should say—he was able to provide to me. In due course, the Forest Service paid me a consultant's fee to research and write my book, plus transportation expenses and a per diem. This subsidy was far more than I could possibly have hoped for, but, still, I like to think that my viewpoint would have been the same without it.

Subsequently, I made a series of trips around the country exploring forests and interviewing forest people. They introduced me, through lessons in the field, to taxonomy, dendrology, silviculture, hydrology, and wildlife management. I called at byroads and backcountry because that was where the forests were. Wherever I went I saw wonderful wild gardens, scarcely known beyond the local communities. I became acquainted with rain forests of the Pacific Northwest and desert gardens in the Southwest. I recognized that timber had a place, but when I looked at a forest I saw trees and the message that comes with them. In choosing the title of the book, I turned to Robert Frost's poem about stopping by the woods on a snowy evening. I wrote:

> Go forth from the house in the village. Watch your woods fill up with snow the darkest evening of the year. Or with the glory of the sunlight of the brightest morning. Listen to the sweep of easy wind and the faint fall of flakes. Listen to the many voices of the forest, the soft, serene, the violent, and natural sounds we sometimes hear but cannot understand. Let us share the promise and the joy, each in his own way, of the good and sweet earth, the woods and the lake.[1]

I found drama, personality, and even romance in forestry and forest history, relative to national parks as well. Study showed that the nineteenth century was marked by wasteful exploitation and devastating fires, and that timber, mining, and railroad interests were bent on carving up America and exploiting its natural resources. In response to rising scientific and public concern, Congress, in 1864, ceded Yosemite Valley to California for safekeeping. In 1872, Yellowstone was set aside as the first national park. The foundation was laid not only for additional national parks but also for forest protection. This led to the Forest Reserve Act of 1891, authorizing the president to withdraw portions of the public domain from disposal and designating them as forest reserves, and ultimately to the Organic Act of 1897, assigning the forest reserves with the mission of protecting watersheds and "assuring a continuous supply of timber for the use and necessity of citizens of the United States." It was the beginning of a structured system of public lands, reaching a new level in 1905, when President Theodore Roosevelt; Gifford Pinchot, his close ally; and their conservation supporters succeeded in placing 180 million acres of forests, plains, and mountains under protection of the Forest Service.[2]

Inevitably, the trail led me to John Muir, who reinforced my inner feeling that appreciation of trees, like love, ambition, and literature, is an emotional and spiritual individual experience. Once he climbed a tree in the heart of California's forestland. It was in the midst of a windstorm and he described it so:

> But the winds go to every tree, fingering every leaf and branch and furrowed bole; not one is forgotten; the Mountain Pine towering with outstretched arms on the rugged buttresses of the icy peaks, the lowliest and most retiring tenant of the dells; they seek and find them all, caressing them tenderly, bending them in lusty exercise, stimulating their growth, plucking off a leaf or limb as required, or removing an entire tree or grove, now whispering and cooing through the branches like a sleepy child, now roaring like the ocean; the winds blessing the forests, the forests the winds, with ineffable beauty and harmony as the sure result. . . .
>
> We all travel the milky way together, trees and men; but it never occurred to me until this storm-day, while swinging in the wind, that trees are travelers, in the ordinary sense. They make many journeys, not extensive ones, it is true; but our own little journeys, away and back again, are only little more than tree wavings—many of them not so much.[3]

Muir in time became a published writer, widely read and recognized, but definitely pursuing his singular style of expression and advocacy. He sought to give voice to God's temples on earth—the mountains, forests, and sacred groves. His series in San Francisco's *Daily Evening Bulletin* in the years 1874–1875 (reissued more than a century later as *John Muir Summering in the Sierra*) was like a series of letters addressed to civilization, in which he asked his readers to see, hear, and believe in a radically new way. In his foreword to *Summering in the Sierra*, Robert Engberg wrote: "Muir invented a new ethics-centered type of reporting, 'wilderness journalism,' to spread his message. Composed in the field and sent to the publisher without later revisions and redrafting, it retains the freshness and spontaneity that characterizes Muir's best writing. . . . Behind every report about people and places lies Muir's own story of why he quit his solitary wanderings to become the leader of the American conservation movement."[4]

In due course, Robert Underwood Johnson, editor of *Century* magazine, urged Muir to write about Yosemite and, largely through their collaborative efforts, Yosemite National Park was established in 1890. Muir's writing flowed into print continually throughout the rest of his life, but was interrupted by political battles, often successful, ultimately leading to his effort to save Hetch Hetchy Valley from the proposed dam across the Tuolumne River in Yosemite National Park.

Gifford Pinchot and Muir met in California in the summer of 1896 and took to each other at once. The former was serving on a federal commission studying the future of forest reserves, while Muir came along to observe. He was in his late fifties, already a legendary and esteemed Western figure and an advocate of the forest reserves, having written:

> Nearly all our forests in the West are on mountains and cover and protect the fountains of the rivers. They are being more and more deeply invaded and, of course, fires are multiplied; five to ten times as much timber is burned as is used, to say nothing of the waste of lowlands by destructive floods. As sheep advance, flowers, vegetation, grass, soil, plenty and poetry vanish.[5]

The study commission proceeded south through California and across Arizona to the rim of the Grand Canyon. Muir and Pinchot by now were inseparable. As the latter recorded:

> With John Muir I spent an unforgettable day on the rim of the prodigious chasm, letting it soak in.... And when we came across a tarantula he wouldn't let me kill it. He said it had as much right there as we did.
>
> Muir was a storyteller in a million. For weeks I had been trying to make him tell me the tale of his adventure with a dog and an Alaskan glacier, afterward printed under the title of *Stickeen*. If I could get him alone at a campfire—we had left from our lunches a hard-boiled egg and one small sandwich apiece, and water enough in our canteens. Why go back to the hotel?
>
> That, it developed, suited Muir as much as it did me. So we made our beds of Cedar boughs in a thick stand that kept the wind away, and there he talked until midnight. It was such an evening as I have never had before or since.[6]

Reading of their intimate dialogue and the upbeat rendezvous between them, one might expect a lifelong Muir-Pinchot camaraderie to follow, with vigorous joint crusading for public parks and forests. But these two forceful men were destined to journey on different pathways, contradictory and ultimately conflicting.

Pinchot, for his part, introduced forestry to America. That was his cause. With family support he established a school of forestry at Yale

University, his alma mater, with others to follow at other institutions, mostly land-grant agriculture schools. He and Theodore Roosevelt already were close friends; thus he became a presidential confidant and adviser virtually from the day in September 1901 when Roosevelt assumed the presidency (on the assassination of William McKinley).[7]

Presidents Benjamin Harrison and Grover Cleveland had set aside a total of 33 million acres as forest reserves. Now, over the next six years, Roosevelt boldly used his authority, in defiance of political odds, to convert 132 million acres from the public domain into forest reserves. His "executive usurpation" was roundly denounced by western senators, but he was undeterred. In 1905, he pushed through legislation transferring the reserves from the Department of the Interior to the Department of Agriculture, renaming them with a stamp of permanence as national forests, with his friend Pinchot as the first chief of the U.S. Forest Service.

John Muir followed a different trail. In 1892 he joined with other mountain enthusiasts in forming the Sierra Club and became its president. In its beginning it was essentially a Northern California group supporting the new forest reserves and national parks.

Roosevelt was a friend and supporter of both Muir and Pinchot. He was an easterner and Harvard scholar who knew and loved the outdoors and the West. "The preservation of forests and wildlife go hand in hand," he wrote. "He who works for one works for the other." And so, in 1908, Roosevelt considered a proposal to proclaim the establishment of a national monument along Redwood Creek in Marin County, California. It was a time when much of the great coastal redwood forest was falling under the ax. This particular fragment was relatively inaccessible and thus far had been spared. William Kent, a prominent and wealthy Republican congressman, interceded while time allowed and purchased the property with his own money. Then he learned he could best save it through a provision of law allowing the federal government to accept lands with historic or other public interest from private individuals. Roosevelt declared it a national monument and named it after Kent, but the congressman insisted on the name by which it is now known and cherished, Muir Woods National Monument. The visitor today will find, at a prominent location, a bronze, life-size statue of Gifford Pinchot.

In 1908 President Roosevelt convened the Conservation Congress of Governors, a historic event attended by all the governors and many other important Americans. The central focus, however, was on the principle of "wise use," as enunciated by the president's ally, Gifford Pinchot. But J. Horace McFarland of Pennsylvania spoke up boldly for the preservation of natural beauty. He said the scenic value of all the national domain remaining should be jealously guarded and that "national parks, all too few in number, ought to be absolutely inviolate."

Then came the proposal to construct a reservoir in the Hetch Hetchy Valley of Yosemite National Park for the benefit of the city of San Francisco. It was an issue bitterly contested through the years of Roosevelt's presidency and beyond. A principal proponent in alliance with San Francisco was Pinchot, Roosevelt's friend and adviser, who argued that it was a means of undermining the privatization of resources. He wanted San Francisco to publicly control its water as a matter of public policy. The Hetch Hetchy dam site would do that.

When the project was finally and unalterably approved in 1913, Muir was devastated, for the issue to him was not simply the "Hetch Hetchy steal," as it came to be known, but the integrity of all the national parks. "We may lose this particular fight," he declared, "but truth and right must prevail at last."[8] Unfortunately, the Hetch Hetchy precedent—compromising the integrity of the national parks—has been repeated many times. That is the tragic fact of parks history.

McFarland was part of a new wave of concern that joined and supported Muir, and carried on after Muir died in 1914. Even without Muir, the national parks movement grew. When President William Howard Taft, in 1912, urged Congress to place a new bureau in charge of the parks, he said it was "essential to the proper management of those wonderful manifestations of nature, so startling and so beautiful that everyone recognizes the obligations of the government to preserve them for the edification and recreation of the people."[9] Three years later, in 1915, Secretary of the Interior Franklin K. Lane sought to unify park management and appointed Stephen T. Mather as his assistant in charge of doing so.

Meanwhile, the Forest Service, in its early years of stewardship, gave the mountains back their dignity. While in the West national forests were formed of land already in public ownership, forests in the East were based on purchase, mostly of land that nobody wanted, with some primeval tracts still surviving here and there. The early foresters were community people. Many hadn't been to forestry school so they didn't know the technical jargon that separates professionals from ordinary folk. The Forest Service achieved its reputation in those early years as a custodian, a protector of the public estate. But over the years the agency and its personnel changed. They became professionalized and institutionalized. Instead of custodians, they became managers, technical people, trained in commodity production, like agronomists, to see trees as timber, awaiting harvest, with scant attention to philosophy or land ethic.

Nevertheless, Aldo Leopold was trained in the system and went on from there. A native of Iowa and a graduate of the Yale School of Forestry, he served professionally in New Mexico national forests. In looking across the Southwest, he saw wilderness diminished, particularly in animal and bird life. "Driven from their once great range by civilization," he wrote, "the wildlife that was at one time America's most picturesque heritage has found refuge in the national forests."[10] He proposed successfully, in 1924, establishment of the Gila Wilderness area, embracing the hazy blue Mogollon Mountains above the Mexican border. That became the model and inspiration for the Wilderness Act, passed by Congress in 1964. Leopold went on to teach at the University of Wisconsin and to define a land ethic in his classic work, *A Sand County Almanac*, published after his death in 1948. Consciously or otherwise, he was echoing Muir when he wrote, "We can be ethical only in relation to something we can see, feel, understand, love, or otherwise have faith in."[11] Both Muir and Leopold certainly convinced me.

During my period of close association with the Forest Service, I was asked to write the text of a booklet, "Trees of the Forest: Their Beauty and Use." Actually, I loved looking at trees in the forest and learning that each of them was composed of root, stem, bark, branch, bud, leaf, flower, and fruit, but I confess that I also looked for the spirit of the tree. Considering it breathed, drank water, nourished itself, and transmitted qualities of heredity through reproduction, it seemed logical to believe there would be a spirit too. But my forester friends endeavored to set me straight, explaining that the tree had no nervous system; it couldn't move to water or shelter, and that, unlike humans and other animal species, it reacted unconsciously to stimulus. They were certain the tree couldn't have a spirit. To most foresters, that tree was meant for use—and for them, that was the beauty in it. But as Muir wrote, "We all travel the milky way together, trees and men."

16

On Becoming a Columnist

My book *Whose Woods These Are: The Story of the National Forests* was well received and well reviewed when it appeared in 1962. I became better known, both as writer and conservationist, and was offered new assignments, including the chance to express my own views as a columnist, first in *American Forests*, then in *Field & Stream*, *Defenders of Wildlife*, the *Los Angeles Times*, and *Woman's Day*. In my column "On the Go," in *Woman's Day*, I continued as a travel writer, trying always to raise the sights of my readers. With the others, it was all conservation and the environment. I tried to do my homework, dig to the truth and stick to the truth, while letting the chips fall where they might. This led people with special interests to complain to one editor or another. I didn't mean to write mean things and upset these people, and wear out my welcome. I enjoyed this work, though it wasn't always fun.

I began contributing to *American Forests* in January 1960. The magazine was the official publication of the American Forestry Association, an old-line organization headquartered in downtown Washington, D.C., at the conservative end of the conservation movement. Most of its directors came from the timber industry, Forest Service, and forestry school faculties. Jim Craig, the editor, was in his early fifties; he had been there for years after earlier work on newspapers. *American Forests* had a substantial circulation for a conservation publication—about eighty thousand at its high point—and Craig tried to keep a balance to it. He told me that he wanted to edit the magazine as a forum open to diverse ideas. That was why he hired me: to "ward off blandness."

At first I wrote rather gently; my first column, in March 1966, was about the Lüneburger Heide, a German nature reserve on sandy heathland that I thought might serve as a model for consideration in America. Then in following columns I reviewed my visit to the forestry school that Gifford Pinchot attended at Nancy in France, and citizen efforts to save Sunfish Pond in the Kittatinny Mountains of New Jersey from a dreadful water storage project.

From the outset, readers sent me tips, confidential government and industry reports, and appeals for help regarding one eco-calamity or another. The readers were largely everyday Americans who liked trees and forests but, among them, there were also foresters, forest educators, and forest industry professionals tending to the store they felt was theirs.

Early on, in the issue of May 1966, I wrote a column about a critical national park issue—the threat of new dams at the borders of the Grand Canyon. That column came about when *Reader's Digest* announced that month's issue would feature a major article by a staff editor, James Nathan Miller, on the proposed dams. But there was more to it: the *Digest* would fly a planeload of media people, myself included, from New York to the South Rim of the Canyon for a forum on the issue.

Looking back on it all, that forum, the charter flight, and probably even the *Digest* article reflected the ingenuity and energy of David Brower, the executive director of the Sierra Club. I had met Brower a few times, but had not seen him in action until now at the Grand Canyon. In this period of the 1960s, Brower was a singular and influential personality in the conservation movement. He was tall—six-foot-three or four—good looking, erect and athletic from his days in mountain climbing and prematurely white-haired. He had attended the University of California for four years without graduating, then worked as an editor for the University of California Press and in public relations for the concessionaire in Yosemite National Park.

When he became the Sierra Club's first full-time paid executive, he transformed the club from a mostly California group of polite outdoors enthusiasts into a militant national organization committed to advocacy. He proved that he was a genius in communication. He was the archdruid, who time and again dared to challenge the impossible, and who became the center of the best-selling nonfiction book *Encounters with the Archdruid* (1971), written by John McPhee, a staff writer of the *New Yorker*.

Now, at the Grand Canyon, he transformed what was billed as a forum into an all-out offensive. Representative Morris K. Udall, of Arizona, who shared the platform with Brower, presumably to defend the dam proposal, was clearly outflanked and befuddled. He looked back on the encounter in his book *Too Funny to Be President*: "The weakness of the arguments in favor of the dams was borne home to me the day I had to debate David Brower—as clever, tough, and tenacious an opponent as you could want—in front of a gaggle of national press at the worst possible venue, the rim of the Grand Canyon." Udall was then a young Arizona congressman doing what was necessary to stay in office. "This was a tough assignment," his narrative continued, "comparable to debating the merits of chastity in Hugh Hefner's hot tub in front of an audience of centerfold models, and me being on the side of abstinence."[1]

The idea was to build two dams, neither actually in the national park but both impinging on it. It followed the dispute over Dinosaur National Monument, described earlier in this book. While Dinosaur was saved, the Bureau of Reclamation succeeded in constructing Glen Canyon Dam on public land outside the park, and in submerging a wild desert of sandstone cliffs and beautiful side canyons beneath a reservoir called Lake Powell. Then, once the floodgates were closed at Glen Canyon in 1963, the engineers turned their focus on the Grand Canyon.

The upstream dam would have been sixty miles below Glen Canyon Dam but inside Marble Canyon National Monument; the other was to be 200 miles downstream at Bridge Canyon, near the point where the Colorado River flows into Lake Mead—both together flooding 146 miles of the Colorado River and the bottom 600 feet of much of the Grand Canyon. It was an engineer's delight, but otherwise a nightmare. It was ultimately defeated, with most of the land in question added to the national park.

While at the Grand Canyon, I spent collegial time with Martin Litton, who was then the travel editor of *Sunset* magazine and a close ally of Brower. In 1952 while working for the *Los Angeles Times*, he wrote a series of articles critical of the dams proposed in Dinosaur National Monument. Brower read that series, was greatly impressed, and invited Litton to join the Sierra Club board.

Litton told me the Sierra Club board wasn't quite sure that it wanted to oppose the dams. "The knowledge of the board members was so fuzzy and thin," he said. "They didn't even know where the Grand Canyon is."

On June 9, 1966, Brower published a full-page advertisement in the Sierra Club's name in the *New York Times* and the *Washington Post.* It was written in commanding and eloquent language, urging public opposition. One day later, the Internal Revenue Service warned the Sierra Club that its tax-exempt status was in jeopardy for lobbying. Brower defiantly placed another advertisement headlined "Should We Also Flood the Sistine Chapel So Tourists Can Get Nearer the Ceiling!"

Saving the Grand Canyon was an absolutely monumental achievement. The secretary of the Interior of the time, Stewart L. Udall (older brother of Representative Morris K. Udall), was one of the conservation heroes of the period, but nowhere near infallible. He was a principal advocate of the proposed dams and later said, "The Sierra Club didn't save the Grand Canyon, the American people did." Maybe so, but it was Brower who lit the torch and mobilized America.[2]

I was still listening and learning. At the conclusion of my column, I wrote:

> At 5:30 a.m. I walked to the rim to observe the sun rise. The entire canyon was in darkness, and so was the day except for a thin band of white on the horizon. From one second to the next the scene changed and the world brightened. This was Arizona, the arid place, wide, wild and wonderful, which Barry Goldwater and Morris Udall extolled for its natural beauty and extolled and pledged to defend to the bitter end.
>
> If I were an Arizonan, I would ask, "Why do we really need these dams? What effect will they have on the values we treasure?" Everybody in the arid country wants dams, or thinks he wants dams, like a panacea, or religion, that is said to guarantee a better life, inducing prosperity, boom, bank deposits, building, freeways, congestion and sprawl. Arizona, I fear, demands the right to make the same mistakes of planlessness that every other place has made. . . .
>
> If I were an Arizonan, I would go up to the Canyon at daybreak and feel that I held its future in my hands. Harnessing natural features and improving upon rivers represents man's splendid ingenuity, beyond a doubt. The greater ingenuity is to know that man survives while leaving a few features of the earth to their own device.

In another early column in *American Forests*, I wrote a tribute to Edward Meeman, a noted newspaper editor and conservationist who had died in Memphis at age seventy-seven. As editor of the *Knoxville News Sentinel* in the 1920s, he strongly supported the move to establish Great Smoky Mountains National Park, and then, in Memphis, he encouraged the entire Scripps Howard chain and other media to conservation advocacy.

My column led to several letters from members of a group called the Citizens to Preserve Overton Park in Memphis. They were touched to read the tribute to their friend Meeman and missed him in the fight to save Overton Park from the construction of an interstate highway through the heart of the park. They felt that even though he was editor emeritus of the *Memphis Press-Scimitar*, Meeman's hands were tied, and that Scripps Howard in Memphis was in league with downtown commercial interests in support of the freeway. They wrote:

> Please help us save this midtown park. The park's 300 acres has trees 350 years old and it is sorely needed as high-rises go up in this area and an old people's home is on one side. Why take federal funds to build an unneeded freeway through a park, then take 50 percent to buy open space (having destroyed a natural, existing one), then use 75 percent to fight air pollution which the freeway has helped bring to the city. How wasteful can we get?

I investigated and found that Scripps Howard (which controlled both Memphis newspapers) had ridiculed the park defenders and tried to create a public impression that the issue had been settled irrevocably and that nothing further could be done by public interest. It belittled any politician who dared to challenge the proposed park invasion. In March 1968 I devoted a column to Overton Park, writing that "an established park is an integral and sacred part of the American city—when the choice must be made between open space and a highway it makes more sense to locate the highway elsewhere than to invade the park."

The Citizens to Preserve Overton Park asked me if there was some agency in Washington, public or private, to help them, since federal money was involved. But almost everyone declined, saying simply, "But we cannot interfere unless the political climate is favorable and a local political body asks us to." Nevertheless, Kenny Dale, who worked in the park practice program of the National Park Service, interceded. That park practice program goes back to the days when Stephen T. Mather was the National Park Service director, when national parks were viewed as the core of a network embracing state and local parks. It doesn't exist any longer; now it's all about national parks and the politics they serve.

In any event, Dale went to Memphis and gave this report:

> The transcendent value of Overton Park for park purposes should be obvious to anyone. Less perhaps are the irreplaceable values inherent in the wooded portion of the park. My observations lead me to believe that the potential recreation and nature education opportunities available in this portion of Overton Park have not begun to be realized. . . .
>
> In my opinion, it is a mistake to use this park for highway purposes, and I am sorry that I can do nothing to prevent it. . . . It is not likely that a city administration that permits the destruction of such values would make much effort to provide them again.

His statement came at a critical time; it gave heart to the park defenders and pause to the road builders. The Overton Park case ultimately was debated in Congress and before the Supreme Court, which ruled in favor of the Citizens to Preserve Overton Park. That park remains an asset to Memphis. Ironically, the *Memphis Press-Scimitar* bet on the wrong horse and folded, while Scripps Howard blithely continued to present an annual award for conservation writing in memorial tribute to Ed Meeman. All this and more came to me through the column I wrote in *American Forests*.

17

Speaking at Yale

I mentioned earlier that I received invitations not only to write but to speak at workshops, seminars, and grassroots rallies. One week I spoke in Milwaukee to forest supervisors of the Eastern Region of the Forest Service and a week later at the Mather Training Center (named for Stephen T. Mather, the first director of the National Park Service), at Harpers Ferry, West Virginia.

I was invited and went to Harpers Ferry four or five times. I suppose that I was identified as a critic, but Raymond Nelson, the director of the training center, was an independent thinker himself. He wanted to stimulate and broaden personnel with challenge rather than with superficial stroking. That worked for a time, until he was transferred to the Washington office, where he was buried in paperwork, leading him to retire to lead the simple life in Maine.

While I met at these training sessions with professional personnel at various levels in the ranks, I found most of them cautious about speaking up, reflecting the power of the peer group in a structured institution. Any independent, irreverent ideas they kept to themselves. New ideas worth talking about came from above, inside the system, and anyone communicating with outsiders like me did so at his or her own risk.

I told them at Harpers Ferry that they ought to be proud of a young backcountry ranger named Edward Abbey, who had lately published a provocative literary work called *Desert Solitaire: A Season in the Wilderness*, and ought to emulate his honesty of expression about fundamental park values. Many of them, especially in the lower ranks, read that book, but they didn't talk about it, especially not at training programs. In due course, Abbey left the Park Service to follow his creative star.

After Ray Nelson's departure, someone in the Park Service told me the study room at Harpers Ferry was rigged so that superiors could record the discussion and determine who said what. That didn't make sense until one day at Capitol Reef National Park, in Utah, the superintendent wanted to share confidential information with me, but he didn't feel safe in his own room and led me to the men's toilet, where he felt he could talk freely.

In this time period I went to speak at the School of Forestry at Yale University. This was a new and different experience for me; Yale was one of the first of various universities to invite me to come and inspire students and faculty. Almost always I was invited by the school of forestry or natural resources, or the department of wildlife or ecology. It was strange, in a way, that deans and department heads would have me talk to their students when, at that time, I did not have a degree to call my own.

I arrived in New Haven the evening before my engagements at Yale. I remember the date clearly: it was April 4, 1968, the very day that Martin Luther King Jr. was assassinated in Memphis. Black ghettos erupted in fury and cities burned. I felt disturbed and depressed by radio and television accounts of violence and fire in the inner cities of the country. I kept waking up during the night, thinking again and again that, in modern America, the educated and affluent escape crowds, concrete, and crime, breathing cleaner air in a cleaner environment, while the poor, especially the nonwhite, are trapped, disenfranchised from the bounties of our time—the lower the income, the lower the quality of life. The country then was deeply divided over the war in Vietnam, a military misadventure that, to me, legitimized mass killing and killing without cause and brought violence rather than tolerance to America. I felt deeply disturbed; I tried to equate my actions and personal goals with the tragedy and meaning of that day and thought about how to make a message of meaning out of it for the Yale students.

The next day, I spoke in classes and then to an assembly of students and faculty at Sage Hall, the headquarters of the graduate forestry school that Gifford Pinchot and his family established in 1900. Pinchot was a Yale man himself, but he studied forestry abroad because there were no schools for it in America at the time. It seemed appropriate to start my remarks by quoting the prominent Yale ecologist Paul Sears: "Conservation is a point of view involved with the concept of freedom, human dignity, and the American spirit." Pinchot expressed the same idea: "The rightful use and purpose of our natural resources," he said, "is to make all the people strong and well, able and wise, well-taught, well-fed . . . full of knowledge and initiative, with equal opportunity for all and special privilege for none."[1] He conceived forestry as the vanguard of a public crusade against control of government by big business. Under his leadership, the Forest Service achieved an early reputation for fearlessness in a system then, as now, afflicted with bureaucracy and timidity.

I told the Yale audience that I had learned an important lesson in Memphis. I had been there before Dr. King's death and had written about the conservation efforts of the Citizens to Preserve Overton Park. On the face of it, the Citizens had nothing in common with the humble black garbage workers whose cause Dr. King had come to defend. Or perhaps they did, considering they were fighting exactly the same economic and political forces.

When downtown merchants and developers decided that a freeway through Overton Park would jingle coins in their pockets, the distinctive urban forest became expendable. A former mayor of the city, Watkins Overton, great-grandson of the man for whom the park was named, courageously spoke of the park as hallowed ground—a priceless possession of the people, beyond commercial value. Nevertheless, he and the upper side of Memphis learned painfully, along with the garbage workers, that democracy can be "a government of bullies." As Overton said, "Entrenched bureaucracy disdains the voice of the people but eventually the people will be heard."

That idea was paramount in my mind. Entrenched bureaucracy of all kinds disdains the voice of the people. It is the weakness of institutions, whether private or public, profit-making or eleemosynary, academic or professional. Institutions, by their nature, tend to breed conformity and compliance; the older and larger it becomes, the less vision the institution expresses or tolerates. But eventually the people will be heard, as evidenced in the ultimately successful efforts of both the garbage workers and the defenders of Overton Park.

I reminded my audience that Pinchot stressed the cause of forestry education to train professionals in a social movement, but I found foresters as technical people, focused mostly on wood production, trained to see trees as board feet of timber. Nor could I say the National Park Service was much different than their Forest Service colleagues. Its personnel might voice concern for ecology as a principle, but scarcely as something practical in critical need of defense. The best defense, at least in my view, is an alert and involved public. But national parks personnel are generally inward-oriented and poor communicators. They know the public as visitor numbers, but not as decision-makers. Woe unto the parks person who goes to the public with faith or trust in his or her heart. The parks person is a "professional," which is how he or she learned to appreciate the values of ecology in theory, but conformity and compromise in practice.

I said that students in most academic programs are bred to be partners of the system, not to challenge it. This is part of the nature of institutions in our time. Whether the issue be social justice, peace, public health, poverty, or the environment, all make candidates for study, research, statistics, coursework, documentation, literature, and professional careers, while the poor remain impoverished, environmental quality worsens, and our last remaining shreds of wild, original America are placed in increasing peril.

Martin Luther King Jr. saw three major evils—racism, poverty, and militarism—and found them integrally linked, one with the other. But the degraded environment is a fourth major evil, also joined with the others. Environmentalists speak of concern with forests, water, air, soil, fish and wildlife, land use, and use of resources, but these are only symptoms of a sick society that needs to deal more fundamentally with itself.

Presidents and congresses, one after another, Democratic as well as Republican, have opposed anything but the most niggardly expenditures to educate and house the poor, provide for the aging, and rehabilitate the imprisoned and the mentally ill; in the very same fashion, they cannot find funds to protect the soil, safeguard the wilderness, or enhance wildlife. The United States has spent vast sums for so-called security from other nations, while for a fraction of that amount it could have extended humanitarian aid and eliminated the threat of war.

These official actions reflect a system that places a low priority on human values and natural values, a system that needs to reorder priorities while there is still the chance. Pinchot may have said it best in 1910, when forestry was a vital, progressive force in the forefront of the conservation crusade:

> We have allowed the great corporations to occupy with their own men the strategic points in business, in social, and in political life. It is our fault more than theirs. We have allowed it when we could have stopped it. Too often we have seemed to forget that a man in public life can no more serve both the special interests and the people than he can serve God and Mammon. There is no reason why the American people should not take into their hands again the full political power which is theirs by right, and which they exercised before the special interests began to nullify the will of the majority.[2]

The priority item on the agenda, as I see it, is for those who hope to heal the earth to join with those who hope to heal the souls of our fellows to bring something new to bear. We must face the twenty-first century with new emphasis on human care and concern. The children of the poor will become rich for what is saved; the children of the rich will be impoverished for what is not saved. It takes considerable courage to stand up against money and the power of politics and institutions. It takes wisdom, or at least knowledge and courage, to work through the system.

When we look at the revolutionary task of reordering priorities, and the sheer power of entrenched, interlocked institutions, the challenge may seem utterly impossible. Yet, individuals working together, or even alone, at the grassroots of America have worked miracles. The odds in Selma and Montgomery, Alabama, also looked impossible; so too in the long fight for the Wilderness Act, and for Overton Park and many other places like it. "A nation that continues year after year to spend more money on military defense than on programs of social uplift is approaching spiritual death," wrote Martin Luther King Jr. who embodied in his own self the challenge to spiritual life. I told my Yale audience that each individual must realize the power of his and her own life and never sell it short. In setting the agenda for tomorrow, miracles large and small are within reach.

Pinchot, in his book *The Fight for Conservation*, warned:

> The vast possibilities of our great future will become realities only if we make ourselves, in a sense, responsible for that future. The planned and orderly development and conservation of our natural resources is the first duty of the United States. It is the only form of insurance that will certainly protect us against the disasters that lack of foresight has in the past repeatedly brought down on nations since passed away. . . .
>
> A nation deprived of liberty may win it, a nation divided may reunite, but a nation whose natural resources are destroyed must inevitably pay the penalty of poverty, degradation, and decay.[3]

The role of government should be to support conservation, physical fitness, and healthy outdoor leisure away from a mechanized, supercivilized world. A wholesome natural environment provides the foundation for a wholesome human environment. We can't have one without the other. The preservation of nature is a use in its own right—a "wise use." That was the essence of my presentation at Yale, and virtually everything I've said or written.

18

The Timid, the Hesitant, the Compromisers Have Failed

Soon after Conrad L. Wirth announced his plan to retire, in January 1964, as director of the National Park Service, the editor of *American Forests* sent me to interview him. I was to invite him to reflect on his thirty-five years in government, all with the Park Service, including twelve years as director, and to ask why he chose this particular time to step down.

I felt that I knew Wirth well, having been with him many times, out in the parks and in Washington. I recognized that his professional career went back to 1931, when he joined the Park Service, but that his experience went back even further, to when he graduated with a degree in landscape architecture from the University of Massachusetts, and to when his father, Theodore Wirth, had been director of parks in Minneapolis.

Wirth recalled that, early on in his career, he met Horace Albright, then the Park Service director, who assigned him to supervise activities of the Civilian Conservation Corps (CCC) in state parks, then a Park Service responsibility. Three years later, Arno B. Cammerer, who succeeded Albright, assigned Wirth to take charge of CCC work in the national parks as well.

We talked about the CCC. In the depression years it became the first "emergency agency" established by President Franklin D. Roosevelt, eventually putting three million men to work in the national parks, national forests, and state parks. By the end of the program in 1942, CCC workers had built scores of bridges, constructed flood-control projects, cut 97,000 miles of fire roads, and planted three billion trees, prompting the nickname "Roosevelt's Tree Army."

I commented that the rustic, rock-and-timber buildings and massive lodges were now famously part of the national parks' visual style, often referred to as "parkitecture." Wirth agreed, adding there was a lot more than CCC work to New Deal emergency and relief measures. Massive sums of money (for the day) were delivered to the National Park Service, multiplying its duties and expanding its sphere of influence. The Public Works Administration (PWA) contributed $47.6 million for land acquisition, roads and trails, parkways, buildings, sewers, and water systems. In connection with forty-six "recreational demonstration areas" near urban population centers, the Park Service received $27 million from the CCC, plus $28 million from the Emergency Relief Administration (ERA) to buy and clear land and build campgrounds, stores, and other service units.

Protection and preservation programs were encouraged and funded, including hiring biologists for the wildlife survey unit. In 1935, Congress passed the Historic Sites Act, making preservation of historic sites a formal national policy. In addition, the Park Service received $2.5 million from the Civil Works Administration (CWA) to conduct a Historic American Buildings Survey—24,000 measured drawings and 26,000 photographs of antique structures considered historically or architecturally important.

"The Park Service even earlier had been closely associated with state park development," Wirth said. "As early as 1921, Stephen Mather was instrumental in establishing the National Conference on State Parks. In the CCC period, many states organized park systems for the first time in order to qualify for federal funds. At their request, we wrote legislation for systems in nineteen states, which had no parks at all or only one or two. During the ten years of CCC, state parks accomplished what they normally would have done in fifty years."

This led Wirth to cite the Park, Parkway, and Recreational-Area Study Act of 1936, which provided for cooperative planning with the states.

> We worked together with no less than forty-three of the forty-eight states. It sparked the state park planning program. There was another aspect of the program under which we acquired marginal lands to develop for recreation purposes. Most were within fifty miles of big cities, like Prince William Forest Park in Virginia and Catoctin Mountain Park in Maryland, part of

> which is now the president's retreat. It was built by the Park Service in 1941 with CCC and special funds for Roosevelt as his famous "Shangri-la"; now, of course, known as Camp David. Forty-eight of these parcels were demonstration areas, purchased by the federal government and then given to the states.
>
> I remember how we were prohibited by law from paying over $10 an acre on the average. To balance out the high cost of land around the cities, we acquired a large tract cheaply on the Little Missouri River near Medora in the Badlands of North Dakota. Ultimately it turned out to be a fine acquisition in its own right, because of the unusual landscape and wildlife, as well as the association with Theodore Roosevelt. It is now the Theodore Roosevelt National Memorial Park [later to be reclassified as a national park].

For me, that was a stroke of genius or sheer good fortune, considering that that national park is both a natural and historic treasure.

As National Park Service director, Wirth faced tough issues: proposals to build dams in some parks and to open others to public hunting; pressure from commercial tourist operators, ski promoters, and road builders demanding greater development of facilities; and pressure from conservation organizations seeking legislative preservation of wilderness.

Wirth was not an ecologist—that term was scarcely known in his day—nor a scientist. He wanted to protect the parks for public use and gave me his interpretation of these issues: "To one person, climbing the Teton peaks is an inspiring experience, while to another driving along the Snake River and looking up at the mountains holds the same appeal. My basic idea has been that parks are for people—for people to use and enjoy, but not with the right to abuse or destroy. The biggest problem has been, and will continue to be, convincing the public of the need for sound management, protection, and preservation."

We talked about his reaction to the administration proposal to establish a new agency, to be called the Bureau of Outdoor Recreation, performing many duties formerly handled by the Park Service. "At first I felt the Park Service should handle recreation planning, as we have done since 1936," he said. "I was wrong. On further study I reached the conclusion that an operating bureau should not handle overall planning. To have an agency of government designed specifically for planning purposes is eminently sound. It has the quality of objectivity that we might lack. Certainly we have no monopoly on recreation and nothing gives me greater pleasure than to see other federal agencies play their part in the picture."

That is what he told me, but whether he meant it is another question. The same goes for his decision to retire at that particular time in 1964. Was it his choice or was it forced? A gloomy pall had been cast over the announcement the preceding October of his decision to step down. It appeared that he had been eased out, or pushed out, by higher officials of the Department of the Interior. While Secretary Stewart L. Udall paid him tribute as "an outstanding public servant," Assistant Secretary John A. Carver Jr. chose the occasion, at the superintendents' conference in Yosemite on October 14, 1963, to pan Wirth for taking tough stands in defense of park values and to excoriate the Park Service in no uncertain terms for a variety of "sins." He said:

> When all else fails, the Park Service seems always to fall back upon mysticism, its own private mystique. Listen to this sentence: "The primary qualification of the Division Chief position, and most of the subordinate positions, is that the employees be imbued with the strong convictions as to the 'rightness' of National Park Service philosophy, policy, and purpose, and who have demonstrated enthusiasm and ability to promote effectively the achievement of National Park Service goals."
>
> This has the mystic, quasi-religious sound of a manual for the Hitler Youth Movement. Such nonsense is simply intolerable. The National Park Service is a bureau of the Department of the Interior, which is a Department of the United States government's executive branch—it isn't a religion, and it should not be thought of as such.

Other circumstances clouded the event. Carver's speech, with its criticism of Wirth, found its way into several newspapers almost as it was being delivered.

His retirement was not forced, Wirth insisted. It had been in his mind a long time in order to make way for younger men. His successor, George B. Hartzog Jr., had been chosen one year prior. These, no doubt, may have been the facts of the case, but questions remained to be answered.

Udall, for example, paid high tribute to Wirth and termed Carver's speech as "the sort of thing you do within the family as indicating you can do a better job in some fields." However, Udall had not yet explained the peculiar manner in which the speech, "within the family," became public—or whether he shared Carver's critical views of national park philosophy.

Wirth summed up our interview with fond words about his old outfit that he had long served: "It is a vigorous, capable, aggressive, and loyal organization, dedicated to serving the public in accordance with the objectives enacted into law by the Congress and the policies established by the administration and the Secretary of the Interior."

That may have been true, but old-timers were still around into the 1950s and 1960s, in very key positions, where they had been since the days of Horace Albright and even back to Mather's time. Certainly I benefitted from knowing people like Tom Vint, in design and construction; Isabelle Story, in publications; Ben Thompson, in new areas studies; and Herb Evison, as director of public information.

In looking back, I feel that Wirth intended to stay until 1966, the fiftieth anniversary of establishment of the National Park Service and the completion of Mission 66. That was not to be his fortune. Once the John F. Kennedy administration took over in 1961, he became a holdover, out of step with the new, young team.

Years later, Jim Faber, Udall's first public information director at Interior, told me, "It took us two or three years to get rid of Wirth, but we did." When I interviewed Udall in Phoenix in 1985, long after he left Interior, he explained why and how:

> Connie wanted to be left alone to run his own show. He was a veteran, older than me, old enough to be my father. He didn't take direction well from the top. . . . He was a good man and a good manager, but when we got into this whole expansion program, a park director had to go to the Cape Cods and the Sleeping Bear Dunes in Michigan and be a politician and help sell the park. Connie couldn't do it. . . . John Carver was horrified. He and Connie were always fighting. Connie didn't like anybody looking over his shoulder, and here we were in this new era. So the minute he hinted he was thinking of going, why I encouraged him and got my own park director. George Hartzog was more my type.

In February 1967 I wrote an article for *Holiday* magazine titled "The Politics of Conservation." The heart of it was in these paragraphs:

> The timid, the hesitant, the compromisers often have failed. The bigger and bolder the program, the greater the chance of success. So I found myself thinking while at the White House watching the President [Lyndon B. Johnson] play the role of the master conservationist. The occasion was a ceremony at which he signed the bill establishing the Cape Lookout National Seashore in North Carolina. "I want to be judged as we judge the great conservationists of yesterday as benefactors of our people and as builders of a more beautiful America," the President told the assemblage. Then he passed out 200 pens. The politicians stood first in line, smiling for the benefit of any reporters present; they were followed by civic boosters from Carolina, leaders of conservation organizations, and the press. When my turn came, I introduced myself. The President looked me in the eye and said, "I'm doing everything I can," as though he really meant it.
>
> I believe the President does mean it, in his own way, but it does not necessarily follow that he is doing everything he can. . . .
>
> Secretary of the Interior Stewart L. Udall during a recent interview was asked about his mistakes or regrets. "All of us," he conceded, "overcompromise and therefore fall short of our ideals." Such overcompromise results from mixing conservation so deeply with political considerations that principle becomes a matter of secondary concern.

As visitor numbers to the parks grew and elected officials realized what they had, politics become primary. They might have cited higher, more noble values, but that isn't what they meant.[1]

19

"Parks Are for People" Makes the Great Society Look Good

When Lyndon B. Johnson became president in 1963 following the assassination of John F. Kennedy, pundits speculated that he would replace Stewart L. Udall as secretary of the Interior. Udall was a third-term congressman, a youthful progressive Arizonan, of Kennedy's generation and outlook, not of Johnson's. He had supported Kennedy over Johnson at the Democratic convention two years before. He was one of Kennedy's young men. Orville Freeman, secretary of Agriculture and former governor of Minnesota, was another. I knew them both. In fact, Freeman and I coauthored a book, *The National Forests of America.* Freeman and Udall were young, personable, and popular, and served until the end of Johnson's term.

In the early 1960s, environmental awareness spread across the land and became a political issue worthy of note by both Democrats and Republicans. Publication of *Silent Spring* by Rachel Carson stirred public concern, leading politicians and public officials to respond to show they cared too. In his 1961 message to Congress, Kennedy declared: "America's health, morals, and culture have long benefited from our National Parks and our fish and wildlife opportunities." As part of the message, he urged Congress to pass a wilderness bill—subsequently adopted in 1964—as well as other legislation leading to protection of Cape Cod, and the forerunner to initiatives protecting other shorelines.[1]

Udall's slogan, "Parks Are for People," dovetailed with Johnson's Great Society. Besides, Johnson liked the politics of national parks. As noted earlier, at times when new parks were established by Congress, he would hold a bill-signing ceremony at the White House. He dispatched one of his own loyal Texas staff members, Charles Boatner, to work as public information chief of the Park Service. Johnson's wife, Lady Bird, became an apostle for a beautiful America and her public relations aide, Elizabeth (Liz) Carpenter, organized media press trips to the parks.

Udall felt the same way as Kennedy. As I knew him, while he was in office and later, he was sincere about it. He meant well and wanted to be identified as a conservation leader. Early in his time as secretary of the Interior, he wrote a popular and influential book, *The Quiet Crisis*, aided by (and with appropriate credit to) several eminent environmental writers, including Alvin Josephy, Sigurd Olson, and Wallace Stegner. They all felt they were privy to access and input, while Udall showed himself sensitive to ethical qualities of natural life and earnest in his desire to contribute positively to them.

However, to use his own term, Udall overcompromised, fell short of his ideals, mixing conservation so deeply with political considerations that principle became a matter of secondary concern. In short, when things got tough, politics transcended principle. On one hand, a stream of slick publications and glowing news releases flowed from Interior during Udall's tenure. They assured America that all was well in the great new Kennedy-Johnson conservation wave, even while Udall was supporting extensive leasing of federal and Indian coal lands and environmentally destructive water projects.

I remember Udall at various phases of his career. I felt that he wanted to be frank and open, but almost every time I came to his office on appointment I could see he was trying to cope with a flood of phone calls and correspondence on his desk. While in office he was the hero of environmental groups and their leaders, as evidenced when they flocked around him at their receptions and dinners at one Washington hotel or another, but that ended when he returned to private life. He told me of the time when both the Wilderness Society and National Audubon advertised the open position of executive director. Though he wrote to express interest in both, he never heard back from either one. When he left government he opened a consulting office in the Mills Building, one block from the White House, but failed to get enough clients to support it over the long run. He also wrote a syndicated column on conservation, for which his partner, Jeff Stansbury, did most of the work, but that didn't last either. He wrote a few books, but not one that might have been titled *Mistakes and Regrets: The Inside Story*.[2]

All secretaries of the Interior, Democrats and Republicans alike, have expressed their fondness for the parks, and willingly turn out for a ribbon cutting at one park or another. But the truth is that national parks are pretty small potatoes in their scheme of things. Of much greater import at the Interior Department are the public domain lands with their hard-rock minerals, petroleum, timber, and forage for cows and sheep; the water that flows out of the public lands (including the parks) to be transformed into energy and irrigation; and the outer continental shelf, the OCS, with its submerged petroleum.

Despite the extensive authority in the office, the position of Interior secretary has never been a steppingstone to anything of consequence—not to the presidency, and not even to the vice presidency. The job does not rate high in the pecking order of the president's cabinet, as compared with secretaries of State and Treasury and the attorney general at the Department of Justice. Nevertheless, it brings with it assorted rewards, challenges, headaches, and hang-ups. There was, for instance, the hassle over President Gerald Ford's first secretary of the Interior, Stan Hathaway. He had such a dreadful prodevelopment, anticonservation record as governor of Wyoming that he was greeted in Washington with far more hoots than hurrahs; after a month on the job he withdrew to Bethesda Naval Hospital to repair his nerves, then headed home to Wyoming.

President Ford, while in college, had worked as a park ranger, so he might have known enough to make a better choice. But that is not how politicians see it. They find the department filled with plums for the picking, while national parks, with their lofty idealistic purpose, lend a mask of credibility to the entire department.

But, after all, the Department of the Interior was established to give land away, not to protect it, and with the giveaways have come scandals and scams, recurrent down to our day. On March 3, 1975, for example, it was revealed that forty-nine officials of the U.S. Geological Survey (USGS) owned stock in companies holding mineral leases on federal lands supervised by the USGS, and that the Geological Survey itself, a major bureau of the Interior Department, had taken no action to force divestiture.[3] Even more recently, the George W. Bush administration at Interior was scarred by the Jack Abramoff probe and scandal, which caught up lawmakers, lobbyists, administration officials, congressional staffers, and businessmen. In 2008, Abramoff was found guilty of defrauding Indian tribes of millions of dollars connected with Indian gaming casinos and of corrupting public officials by trading expensive gifts for political favors.

J. Steven Griles, the deputy secretary of Interior in the Bush administration, was heavily implicated in the Abramoff investigation. He had earlier made a career as a coal, oil, and gas industry consultant and lobbyist; Griles resigned after an investigation revealed conflict of interest involving inappropriate communication with his former clients.

Another scandal came to light in September 2008, when the inspector general of the Interior Department reported that employees accepted gifts from oil and gas companies, participated in "a culture of substance abuse and promiscuity," and considered themselves exempt from federal ethics rules. The report resulted from investigations of more than a dozen current and former employees of the Minerals Management Service (MMS), which collects about $10 billion in royalties annually. It found individuals wholly lacking in government ethical standards and management that practiced passive or purposeful ignorance.

The investigation discovered that nearly a third of the entire staff of MMS's royalty-in-kind (RIK) program socialized with and received a wide array of gifts and gratuities from oil and gas companies with which the agency was conducting official business. The RIK program allows companies to pay royalties in the form of oil and gas rather than cash.

The dollar amount of the gifts was not enormous, but the gifts were accepted with "prodigious frequency." Two RIK marketers received gifts and gratuities on at least 135 occasions from four major oil and gas companies with which they were doing business. The investigation also found drug and sex abuse both inside the program and "in consort with industry." One supervisor engaged in illegal drug use and had sexual relations with subordinates. Several staff members admitted to illegal drug use as well as illicit sexual encounters, and some had sexual relationships with industry contacts. Some employees escaped any actions against them by retiring "with the usual celebratory send-offs." The collusion between MMS and the oil industry became headline news with the massive 2010 BP oil spill in the Gulf of Mexico, after which that agency was completely restructured and renamed.

The most infamous Interior scandal was the Teapot Dome oil give-away of the early 1920s. It was neither the first nor the last, and the worst scams may not even be known. To briefly review Teapot Dome: Controversy over whether government land should be sold outright in response to valid claims or leased and maintained in government ownership led to passage of the Mineral Leasing Act of 1920, providing access to oil, gas, and coal on public lands on payment of a royalty fee. That much was considered progress, though it scarcely eliminated the temptation, weakness, and corruption that crystallized in Teapot Dome.[4]

That misadventure began with the election of Warren G. Harding as president and his appointment of Albert B. Fall, a former senator from New Mexico and poker crony from Harding's days in the Senate, as secretary of the Interior. In those years, there were four naval oil reserves, including Teapot Dome in Wyoming and Elk Hills in California. Fall talked his cabinet colleague, the secretary of the Navy, into giving administrative control of the Navy's reserves to Interior, a move that Harding approved with a secret executive order on May 31, 1921. Fall then leased Teapot Dome in its entirety to Harry F. Sinclair, a petroleum kingpin, secretly and without bids.

Soon after, Fall leased a large part of the Elk Hills reserve to Edward L. Doheny, again without bids. The leases came to light, the Senate investigated, and Fall's public career was finished. Doheny later told of lending Fall $100,000 on an unsecured note while seeking the Elk Hills leases; the money had been delivered to Fall's office at Interior by Doheny's son in what became known as "the little black bag."

Fall left the government and went to work for Sinclair. But that was not the end. Indictments were issued against Fall, Sinclair, and the Dohenys. The courts found the Teapot Dome lease had been obtained through collusion and conspiracy between Fall and Sinclair, and that fraud and corruption were involved in the California case. Doheny was acquitted of charges of bribing Fall. Sinclair was sentenced to six months for contempt of court, while the hapless Fall was sentenced to a year in prison and a $100,000 fine.

Every secretary of the Interior has carried the Teapot Dome burden with him, but none tried harder than Harold L. Ickes to overcome history and build an image of integrity for his department.[5] He was a Chicagoan, not a westerner—a Bull Moose supporter of Theodore

Roosevelt in 1912 who had been a journalist and then a lawyer. Once he came to Interior, Ickes served as secretary from 1933 to 1945, longer than any other person before or since.

Critics said he didn't know one end of a cow from another, but he worked hard at conservation, championed progressive New Deal causes, and left noteworthy accomplishments. Ickes abolished racial segregation in the department's restrooms and cafeterias. He invited Marian Anderson to use the Lincoln Memorial for a concert in 1939 after the Daughters of the American Revolution had refused her the use of Constitution Hall, and he personally introduced her. His appointee as head of the Bureau of Indian Affairs, John Collier, endeavored a heroic and historic change in direction in government treatment of Native Americans.

Ickes, self-styled "America's No. 1 Curmudgeon," or "Sour Puss," devoted considerable energy to the National Park Service. In 1938 he married Jane Dahlman, who worked in the historical division of the agency. (He was sixty-four, she was twenty-five.) More than any other secretary of the Interior before or after his time, he enunciated a clear policy of preservation and committed himself to make it work. At the superintendents' conference in February 1936, he expressed this philosophy on the administration of national parks:

> Our national parks are intended to be breaks on this route of insane life we have led. But we find ourselves speeding up to keep pace with a life that is lived at too high a rate instead of performing the function that I think we were intended to perform—of slowing down and making it possible for the people to relax.
>
> If people are genuinely interested, we ought to satisfy that interest. But let's don't drum up trade. Let's let people go into the park and be natural. Let them go back to nature. And don't let's try to force the issue as to people being educated. I don't think the parks were intended to be classrooms.
>
> Colonel Thomson [Charles Goff Thomson, park superintendent] told me when I was out there [Yosemite] that they were beginning to demand shower baths, and running hot and cold water, and all the rest of it. That is perfectly natural. And that is what I am trying to emphasize. There is no limit to it. If you give them hot and cold running water for shower baths, the next thing they will want will be their breakfasts in bed.[6]

At the Department of the Interior, lofty, high-sounding pledges have been common, but rhetoric and public relations have outrun performance. True, there have always been the pressures, from industry and from politicians speaking on their behalf. For example, where President Dwight D. Eisenhower had turned off the flow of federal dollars to build dams in the West, Kennedy, during the election campaign, promised to turn it back on. Then it came time to deliver. As Udall explained to me, Ted Sorensen of the White House insisted that "we have to keep that commitment to the West." Udall supported the Central Arizona Project because the politics of his home state demanded it for his survival and future.

With seventeen agencies and fifty-five thousand employees under his control, he once told me that it took two and a half years to get on top of his job as secretary. He recognized his assignment as political and tried to keep everyone happy—environmentalists, senators, congressmen, and governors, particularly those from the West. People like Representative Wayne Aspinall of Colorado, chairman of the House Committee on Interior and Insular Affairs, pressed him not to be too hard on the mining industry. The job of secretary of the Interior became a tightrope act. He would make a move founded on strong personal belief, but then someone of importance would demand the contrary.

Congressional westerners have always felt that they owned Interior, the attitude in Congress being, "If you don't interfere in my backyard, I won't interfere in yours." Besides, urban congressmen, particularly easterners, know little about public lands and don't have much to gain by learning about them. Because Kennedy rewarded Udall with appointment as secretary, California Democrats got their candidate, James Carr, named undersecretary, in part, at least, to protect their state's interest in the continuing tussle with Arizona over the diversion of water from the Colorado River.

Congressmen and senators who are close to the winning candidate or hold key committee assignments demand their say in filling jobs above the civil service level. It's a chronically terrible system that gives congressmen improper influence in the executive branch, dividing loyalty and militating against effective government. Udall found that he had to consult Senator Clinton Anderson of New Mexico on almost everything. Anderson would say, "You haven't got anybody from New

Mexico. I want somebody in your cabinet." So Udall had to learn not only how the wheels turn at Interior but how to get along with a team of people not of his own choosing.

Udall kept Floyd Dominy, the aggressive dam builder, whom he knew from Arizona and who was favored by Senator Carl Hayden, as director of the Bureau of Reclamation, but he dumped Conrad L. Wirth as director of the National Park Service. He told me: "I picked George Hartzog as director. I had to finagle Connie Wirth, sort of push him out the door a little bit early. He wanted to pick his own man to succeed him. I said, 'Make a list of five. I'm going to participate. It isn't your choice.' So I picked George, and George ran a helluva show."

Udall accepted compromise and considered it necessary to progress. This was clearly evident in the struggle over the California redwoods. President Johnson's 1964 conservation message proposed eight new national parks, including the redwoods, the longtime objective of citizen conservationists. Though about fifty thousand acres of virgin redwoods were already protected in eighteen California state parks, some of the choicest groves were privately owned and subject to imminent logging. In 1963 the Sierra Club published *The Last Redwoods* as part of its campaign, while Udall, at the club's biennial wilderness conference, pledged his support and the support of the administration in the crusade for the park.

The National Park Service dispatched a team to survey the area. After due study the team strongly recommended that the Redwood Creek watershed, with its abundant great trees, become the core of the new park.[7] The site, however, was owned by three major timber companies, in contrast to another potential site, with only half as much old growth, on Mill Creek, that was owned by a relatively small local outfit. The attainable at Mill Creek, to Udall, became the desirable, manifesting what he called the "art of the possible." He wished, he said, "to pick a park, not a fight," and he recommended an area containing few virgin forests not already protected by state parks. So did Hartzog, who dismissed the Redwood Creek study and recommendation of his own agency as not "professional."[8] That sort of derogation of conscientious performance must make any government employee think twice the next time around. Citizen park advocates clung to the dream of the Redwood Creek watershed, portions of which later were added to the park, although only after some of the best of the trees had been logged.[9]

In 1966 the Department of the Interior announced an agreement with a syndicate of public and private utilities to build the Central Arizona Project (CAP), a network of massive coal-fired steam and electric-generating plants in the Four Corners region of the Southwest, where clean air was a primary asset for many years. Udall, one of the main advocates of this project, pledged it would be an absolute model of pollution control.

This assurance was echoed by his brother, Morris ("Mo"), who served in Congress and was chairman of the House Interior committee. Both were strong, aggressive advocates of the CAP. Mo defended CAP before a review team appointed by the Jimmy Carter administration, saying, "I am convinced that the net impact of this project on man's environment will be one of enhancement," while Stewart claimed the Hopi and Navajo Indians of the region would greatly benefit from exploitation of "their under-utilized coal resources." The welfare of those Native Americans was also a responsibility of the Interior Department. They did not ask for the project; nevertheless, Udall advocated a massive thermal-generating plant fueled by two coal strip-mines from their reservations.

But it never proved to be that way. The skies over the Southwest desert have been clouded, smoggy, and smoky with fly ash, sulfur oxide, and nitrogen oxide and the Native Americans have not benefited at all. In *Mo: The Life and Times of Morris K. Udall*, by two journalism professors, Donald W. Carson and James W. Johnson, the reader learns that the CAP cost almost $3.5 billion more than anticipated, and its victory left an ambiguous legacy. The plant at Page, Arizona, did indeed result in huge strip-mines that scarred the Hopi and Navajo reservations:

> Emissions from the plant created air pollution in the pristine Four Corners area that added to the buildup of greenhouse gasses in the atmosphere. The plant transformed the reservations into an energy colony for the benefit of whites while the Native Americans received a fraction of the market value for their coal and, in exchange for the plant, waived their claim to 50,000 acre-feet of water that had been apportioned to them in 1946. . . .
>
> The CAP spigot was turned on in Tucson in 1992. It delivered CAP water to 85,000 homes, but the tap was turned off in September 1994 because of outcries that the water was so full of minerals it corroded water pipes and poured out bad-tasting,

> foul-smelling, murky water. The city began pumping groundwater again. At the start of the twenty-first century, the city was still looking for a way to use CAP water. Without it, city officials said, the water table would continue to drop, sinkholes would develop, and the quality of water would diminish.[10]

Udall later called CAP "an example of bad planning, an example of bad economics," though he promoted it for fourteen years. In due course he conceded that damming Glen Canyon was a mistake, that it should have been part of a national park, although his brother Mo had defended the dam, calling Lake Powell an "incredibly beautiful lake," saying "new beauty has been created, and now for the first time people can get there to see it."[11] The Udall brothers on the whole meant well but, when the chips were down, they yielded to the politics of power with the natural resources of Arizona and the Southwest.

A fallout was felt in the national parks. Field personnel learned to go along or perish. Interpreters at Mesa Verde National Park learned the hard way. They wanted to tell visitors about the Black Mesa coal project while it was still in the proposal stage, so they would know about the potentially severe impact on Navajo and Hopi peoples, the country's last remaining stronghold of traditional Indian life.

On August 16, 1970, a memorandum that read like a military order was distributed to the interpreters by chief park archeologist Gilbert R. Wenger on behalf of park superintendent Meredith Guillet:

> The superintendent wants the distribution of any literature specifically calling attention to the Black Mesa coal project to be stopped immediately. If interpretive personnel are telling visitors on tours to stop by the museum and pick up such literature these individuals are doing so against direct orders. These orders must be obeyed. I do not want to hear of any more incidents on tours or at campfire programs where the uniformed person removes his badge or covers it and says he is speaking as a private individual on these issues. That individual is still on duty and will be guilty of insubordination.

The following morning four rangers turned in their badges and quit. They went out with a public blast: "Morally we felt we could no longer work for an agency whose purpose is to protect our cultural heritage, but whose practice is censorship of major environmental problems which will ultimately affect the very park in which we were working."[12]

After my article on the politics of conservation appeared in *Holiday*, I received a phone call from the editor of *Field & Stream*, Clare Conley. He said that he was impressed, knew of my work elsewhere, and wanted to give his readers more than tall tales about the old fishing guide named Joe and the blood and gore connected with killing a bear.

I was not a hunter at all, and not much of a fisherman, but that wasn't why he wanted me. After we talked, I wrote two articles for him and he called again, expressing confidence that I could bring something of value to the magazine; he offered to engage me as conservation editor with a monthly column, and I accepted.

With a circulation of almost two million, *Field & Stream* was a much larger magazine than *American Forests*, but I proceeded to write for both. I was fortunate that they were essentially noncompetitive, although I tried to assess the difference in the readerships and provide accordingly. In *Field & Stream*, I focused on showing the effects of logging on wildlife and its habitat. I traveled around the country, investigating and writing about politics, bureaucracy, and corporate power, naming the wrongdoers, trying to involve and activate readers so they would not feel helpless against the odds.

I felt that we—Conley and I—were on the right track, breaking new ground and showing the way for our readers. In both magazines, *American Forests* and *Field & Stream*, I believed I was writing truth to power, as I was meant to do.

Inevitably, in my column in *American Forests*, I turned to the failures of forestry. I did not make up or choose the issues; they were already evident and debated in the pages of the magazine and elsewhere. I felt the Forest Service had moved away from its own legacy. Instead of serving as a land steward, the agency had shifted to treat the forest as a commodity, justifying the construction of logging roads and timber sales in fragile country. And so I wrote in February 1970: "The day when 'the forester knows best' is over. The sooner this is recognized the better off we all shall be. Foresters, with all credit to them, rarely have the depth and breadth of vision to make ecological and environmental judgments."

In a letter to the editor published in the magazine later that year, in August, Leon Minckler, a forestry professor at Virginia Polytechnic Institute, wrote:

> Clear-cutting is not necessary for successful hardwood regeneration and it is not the only way that hardwood forests can be managed. If foresters say it is the only way, they are misleading the public. As professionals, foresters must know the alternatives which will meet the needs and desires of woodland owners and the general public.

My friends in the agency had taught me to ask questions, but I asked questions they could not answer. Foresters didn't like my columns, but other readers did and wanted more. Jim Craig, the editor of *American Forests*, ran my columns as I wrote them and stood behind me as long as he could. Ultimately, after six years, I was fired from that magazine. Many readers and supporters were distressed and protested. One of them, Walter J. Hickel, who followed Udall as secretary of the Interior, wrote, "Mike Frome tells it as it is, not as we like to think it is."

I still had *Field & Stream* and felt unfazed. Paper victories are tough enough to come by, but they create illusions rather than true progress. The American Forestry Association was ultimately exposed as a timber-forestry front and lost many, many members.

20

Building an Empire through Political Patronage

Over the years, I observed an abundance of environmental reforms, but reforms, halfway measures, are what they often proved to be. I saw them celebrated by politicians and public officials and also by leaders of environmental organizations, who hailed and claimed credit for another major victory, when it was a victory mostly on paper. Sierra Club executive director David Brower wanted more, and so, for all his aggressive actions, he upset Sierra Club directors. They disliked his uncompromising challenge of government officials, agencies, and their scientists.

Boards of directors, even of conservation groups, often are composed of reasonable, wholly respectable people. They don't go around looking for fights; they like to attend social functions with high government officials, and they gravitate to the lesser of evils. Brower's directors felt he was emotional, irresponsible, and insubordinate. He took the limelight; they were left out. He published classic books on one endangered place after another, but the books didn't make money, and the directors didn't like that either. He made important commitments without approval of the board, which ultimately chose to depose him.

I felt much as Brower did. In the October 1971 issue of *Field & Stream* I wrote a column titled "Let's Put the Public Back in the Public Lands." It began as follows:

> Ownership of public lands is a relatively easy matter. We the people have held millions of acres as a common estate for lo these many years and are still acquiring more, as we must. But how to insure effective, wise management in our own best

> interests? This can be done only through the enlightenment and involvement of people in the process of decision making, in a free and open society. We cannot leave it to the experts, for the experts have gone astray.
>
> Consider the National Park Service and the United States Forest Service. They are responsible for protecting and perpetuating some of our choicest federal resources, but both are stumbling and fumbling. Over the years I have enjoyed intimate relationships with these agencies in Washington and in the field. But I feel they have lost the sense of commitment to a long-term goal on behalf of the people. The outfits were great in their day, but are now ailing and impoverished of inspiration. This shows up in the lands in their care.
>
> "The great danger is that resource professionalism will become shortsighted in its perception of the public good," wrote Charles Reich [a Yale professor, in *The Greening of America*, 1962]. "It may care so much about today's balance sheet that it forgets about tomorrow's heritage."

There was more to this particular column, but this much shows the heart of the matter. For months following I received a steady flow of mail from park personnel across the country, leading me to write another column, in the March 1972 issue of *Field & Stream*, headlined "The National Park Service at 100: An Empire in Trouble." It began:

> For over twenty years I have enjoyed close relationships with the National Park Service and the people working for it—rangers, naturalists, biologists, historians, and park superintendents, out on the ground as well as officials at headquarters in Washington. Many have shared personal secrets with me, of wins and losses and hopes for years ahead. You might expect, therefore, that I would be standing on my chair leading the cheers for the 1972 National Parks Centennial—marking one hundred years since the establishment of Yellowstone, the first national park in America, first anywhere in the world. But I simply can't work up the enthusiasm to shout, "Great work, fellows! Keep it up!"
>
> It just isn't there this time, at least not for me. The National Park Service has lost its way, its sense of commitment, and its openness before the public. Many of the troops in the ranks are frustrated in their inability to manifest vigor, aggressiveness on the public behalf. The old loyalty and esprit de corps are shot; extremely able people with much to say speak only with trepidation and caution, or not at all. . . .

> National parks are overcrowded, overvisited, and overpolluted, and yet the drive goes on to attract even greater numbers for activities inharmonious with trail and campground country.

That column touched a nerve. It led to yet another one, in the August 1972 issue, titled ". . . for the good of the service," which introduced the subject with this paragraph:

> Since my recent criticisms in this space of Park Service misadventures under its director, George B. Hartzog, Jr., I have been overwhelmed with communications from men in the ranks—rangers, naturalists, and park superintendents—hungry for better ways. In the past few months I have been meeting with many of these men and exchanging correspondence with them. They tell me that life in the parks has changed from a joy to an endless siege of coercion, fear, intimidation, and distrust. These highly motivated public servants deserve to be heard.

Their testimonies, some anonymous and some signed, followed. Here are a few extracts:

> I am not just concerned about myself or my family, or my fellow rangers and naturalists. I am most deeply concerned about the great national parks and the agency that was established to care for them. Many of us who have loved the parks and who believe in the original national park concept see all idealism and meaning being lost. Idealism and dedication are feared by those who cannot comprehend, and today the NPS is led by the uncomprehending. The national park land ethic is dying.

> He is very careful to surround himself with those whose spinal columns are well annealed, making the voice of dissent or a suggestion to alter existing priorities a rare occurrence. This, coupled with an obsession to build an empire through political patronage, has reduced concern for resource conservation and proper maintenance of facilities to near-extinction. The National Park Service as a conservation organization should be placed at the top of the endangered species list.

> We are busy creating new platitudes and giving lip service to lofty goals, but we are actually being molded into hypocrites. The slightest inclination to voice concern for traditional values brands one as a rigid malcontent, not flexible enough to hold a position in the glorious New Thrust.

I was surprised by the cascade of correspondence I received. The letters and phone calls kept coming. They confirmed my own observations and suspicions. I thought of the many friends and fine people I had met in the ranks of the agency, starting with my travels on the Blue Ridge Parkway years before. I knew the first three superintendents, Stanley Abbott, Sam P. Weems, and Granville Liles, and felt they were friends, and all committed to the highest standards of public service. Abbott, for instance, was trained as a landscape architect; he had designed parkways for New York State and later became superintendent of Colonial National Historical Park in Virginia. I recall riding with him and discussing park policies and principles. "I'm a Drury man myself," he said, referring to his support of Newton B. Drury, an early National Park Service director and staunch preservationist. But I doubt that few in the later crowd knew little if anything about Drury or what he stood for.

John McLaughlin could have been a Drury man, too. He started at Yellowstone in 1928, under Superintendent Albright, then rose from a buck ranger up through the ranks to become park superintendent, at Grand Canyon and then Yellowstone, where he had begun. In the 1970s, when an issue arose of coping with an overpopulation of elk in the park, McLaughlin believed the number should be reduced by rangers, as in times past. But local hunters wanted to be invited to get their share, and the politicians supported them. McLaughlin was overruled by Hartzog, replaced, and transferred. His successor, Jack Anderson, made a better fit in the new wave of park management. For one thing, he opened the park to snowmobiles.

I can't say that Hartzog was involved, or consulted, in deciding to open Yellowstone, but this reflected the general feeling in the agency while he was at the helm. The January 1977 issue of the *Courier*, the National Park Service house organ, on its first page, carried the headline, "Crashing through the snow," followed by an extensive feature on the endowments of snowmobiling. It quoted Yellowstone's chief ranger, Harold Estey, who spoke proudly of the park's policy: "Snowmobiles are kind of a natural in this part of the country."

Park Service people lived to rue the day. Later they awoke to the folly at hand and tried to undo it, but once such an intrusive misuse is sanctioned in a national park it is extremely difficult, virtually impossible, to reverse course. The use and number of snowmobiles in Yellowstone have been disputed for years, in and out of the courts. In following one court case, Bill Wade, chair of the Coalition of National Park Service Retirees, said, "This decision reaffirms the most essential value of our national parks—that these are among the most special places in our country where Americans are supposed to be able to enjoy the nation's cleanest air, undisturbed sounds and quiet of nature, and wildlife living as free as possible from the pressures of our modern society."[1] Maureen Finnerty, a retired park superintendent, added, "It is time for the National Park Service to stand up and do what's right. Yellowstone and all national parks were established to provide visitor enjoyment—not recreation using motorized vehicles—but enjoyment of scenery unimpaired by polluted air, wildlife in a natural setting, unmolested by traffic, and tranquility that isn't compromised by ever-present manmade noise."[2]

In this same period of the 1960s and 1970s, Phillip Iversen, superintendent of Glacier National Park in Montana, resisted pressure to open his park to snowmobiles, even when it came from Senator John Melcher. Iversen established a boundary and proved a point; his decision has protected Glacier from snowmobiles ever since.

But Hartzog's influence would be felt here, too. Iversen retired and was followed by Keith Neilson, well respected but definitely of an earlier age. In one sense, he fit well into this great wilderness sanctuary of rugged mountain splendor, one of the last strongholds of grizzly bear, bighorn sheep, mountain goats, moose, and elk. But Glacier had its problems—extensive fires, the negative reaction of private landowners to a proposed ban on water skiing at Lake McDonald, and the aftermath of the deaths of two young women campers killed by grizzly bears in 1967. Besides, Neilson was sixty-four, so he, like McLaughlin, was transferred to a more benign setting.

William J. Briggle, age forty-three, succeeded Neilson as park superintendent in June 1969. He already had twenty years of Park Service experience, in the field and in Washington, and was a protégé of Hartzog. But Dr. James R. Habeck, a University of Montana botany professor who conducted extensive studies in the park, did not see Briggle playing a role in preservation. In a letter published in the *Missoulian* on December 22, 1971, Habeck wrote:

> The pattern of Mr. Briggle's behavior in Glacier Park since 1969 can be better understood when it is revealed that he arrived in Glacier Park after a decade of training and experience in

> recreation areas and recreation planning. It also now makes sense why news releases from Glacier have been captioned, "Briggle Boosts Glacier for Winter Playground" and "Glacier Park Features Variety of Winter Fun." Mr. Briggle did not, of course, write the captions, but the information printed is accurately summarized in these titles. Most of us in western Montana are also familiar with what appears to be an inordinate preoccupation with visitor numbers, and we are told that Glacier has enjoyed a good year—whenever a new visitor attendance record is set.

Then there was the episode at Yosemite, and the downfall of its superintendent. For years, Yosemite Valley, barely seven miles long and a mile and a half wide, was operated like a commercial carnival, strongly influenced by the concessionaire, with abundant elements of urban life, including traffic jams and long lines waiting in restaurants and shops. In this context, a wild counterculture July Fourth celebration in 1971 at Stoneman Meadow seemed almost natural. It turned rangers into cops and was subdued only by force. Hartzog decided that someone in the ranks of his agency had to accept responsibility and be disciplined. That someone was Wayne Cone, park superintendent for less than a year, who was summarily relieved and replaced.

In 2011, during the course of writing this book, I sought the recollection of Doug Mackie, who worked in Yosemite for many years. He wrote to me as follows:

> Wayne Cone was extremely ethical, a walk-around superintendent who got out into the park to stay in touch with the operational issues. I remember once a group of us young rangers were told we were to be furloughed because of budget austerity issues. Mr. Cone gave us our day in court, and began by telling us the first furlough, if such actions were necessary, would be his. Needless to say, end of meeting with respect and satisfaction. He was also the director of the Albright Training Center [at Grand Canyon] when I attended it in 1967. Our entire class had the utmost respect for Mr. Cone. Without trying to cast stones, I believe he took the political rap for the Yosemite riot, a very tragic circumstance for such a fine man.

That is the way things went under Director Hartzog. In the August 1972 issue of *Field & Stream*, I quoted a letter from one Park Service ranger in the field:

> If Park Service employees go to their senators or congressmen, the letters usually filter back to their bosses and they are blackballed. The employees are afraid to say anything because they might get transferred or fired.

I cited the Current Biography Yearbook of 1970, which recorded that "George B. Hartzog, Jr., Director of the National Park Service, is an accomplished public speaker and an adroit political strategist. On Capitol Hill he is reportedly highly respected, although some leading conservationists are said to consider him too much of a 'wheeler-dealer.'"

I wrote that I considered this evaluation accurate but incomplete:

> Mr. Hartzog has insured his own survival through fair weather and foul by catering diligently to holders of key congressional committee chairmanships, notably Senator Alan Bible, Representative Wayne Aspinall and Representative Julia Butler Hansen. Meanwhile, out in the field, his agency is afflicted with one reorganization after another. Personnel are transferred like pawns on a chess table. Good men hardly get acquainted with one assignment before they are told to move to another; families of friends of mine have been uprooted to the point of utter despair.

Hartzog had begun his career as a government lawyer, worked in concessions management and as assistant superintendent in two national parks (Rocky Mountain and Great Smoky Mountains). Then, as superintendent of Jefferson National Expansion Memorial in St. Louis, he was in charge of building a huge stainless steel arch—a monumental marvel, mid-America's answer to world-class structures like the Eiffel Tower in Paris. In August 1962, he left the Park Service to become executive director of Downtown St. Louis, Inc., a kind of chamber of commerce with another name. Shortly, the secretary of the Interior, Stewart L. Udall, whom he had earlier met and impressed, convinced him to quit with a promise to appoint him director of the National Park Service.[3]

Hartzog returned to Washington in 1963, serving as associate director under Wirth and learning the ropes, and then taking over eighteen months later. For those who knew him, Hartzog was breezy and tireless, a listener but never at a loss for words; he was overweight, a hard drinker and heavy smoker, puffing cigarettes in between big cigars, a super salesman adept at cultivating politicians and power.[4]

Hartzog could be charming, disarming, and persuasive, but he was difficult to pin down. His own people were on guard, unable to act because they didn't know what he really wanted. He was careful about the people around him. Through reorganization and reshuffling, the cast of characters became composed of lawyers, planners, and professional managers, with little background in the parks or natural resources. Hartzog could say, "Our national parklands, after almost one hundred years, remain unimpaired for our continuing benefit and enjoyment," and the snowmobiler and skier in Yellowstone, on seeing bison and elk warming themselves around the geysers, would agree.

The old ranger-bred park people were concerned about the advent of a new office-oriented type of "manager." The word went out from above that there were too many rangers and that people with a "broader view" were needed to interpret parks. Those in offices would be classed as "professionals," while those in the outdoors would be "technicians," reduced to the lower federal grade levels, in reduced numbers, with diminished time and reduced training for duties in the backcountry. It was all in keeping with the new order.[5]

Udall and Hartzog moved to establish more parks, bring them closer to people, and make them easier for people to visit. During Hartzog's nine years as director, seventy-eight new areas were added to the national park system. These included national parks of the traditional type, plus historical and archaeological monuments, recreation areas, seashores, lakeshores, riverways, memorials, and cultural units. The first major urban recreation areas, Gateway in New York and Golden Gate in San Francisco, were acquired and established in 1972. The largest development came with passage of the Alaska Native Claims Settlement Act of 1980, adding forty-three million acres to the national park system and millions more to national wildlife refuges and wilderness.

Many individuals and citizen groups outside government played important roles in campaigning for their particular projects. So did members of Congress, for a variety of reasons. Some truly believed in conservation and the place for park preservation, while others endorsed parks as good politics. Yet this massive growth made Hartzog a hero to many personnel in the ranks. He was plainly a savvy political operator and a leader who made them and their work feel important.

Hartzog was dismissed after President Richard Nixon's reelection in 1972. He alleged he was fired for denying Bebe Rebozo, Nixon's close friend, access to a Park Service dock in south Florida. But there is more evidence that he was too independent, too inclined to make end runs around superiors. "George Hartzog can walk through an assistant secretary easier that I can step over a three-legged stool," said Rogers Morton, secretary of the Interior, who was six-foot-seven. From the Nixon standpoint, he was an overripe melon held over from the Kennedy-Johnson administration.[6]

To many, "Big George," until his death in 2008 and beyond, remained the hero who had ushered in the new age. On May 11, 1985, the George B. Hartzog Jr. Visitor Center was dedicated at Jefferson National Expansion Memorial in St. Louis, bearing an inscription: "A forceful and inspiring leader who specialized in crisis and favored bold strategies, Hartzog was one of the great builders of the National Park System."

Reshaping Our National Parks and Their Guardians: The Legacy of George B. Hartzog Jr., a biography by Kathy Mengak, was previewed with praise in the Spring 2011 issue of *National Parks* magazine. The Summer 2011 issue of the magazine that followed, however, included a letter to the editor titled "A polarizing figure." It was written by Gail McLaughlin Stephens, of Frederick, Maryland, who I presently identified as the daughter of John McLaughlin, superintendent of Yellowstone. She wrote that she was disturbed by the review:

> I witnessed firsthand the effects of the other Hartzog. The author notes "the brusque taskmaster with a short fuse." It was brutal.
>
> My father was a longtime member of the Park Service. He began his career as a ranger and worked his way up through the ranks to become superintendent of key national parks. He knew the parks and he knew the Park Service, but when he made a controversial decision, and angered the local politicians, Hartzog would send him off to a park in another state, even though Hartzog had agreed with the decision in the first place. It was very hard on my father, but he rallied and went on, believing that the parks were more important than any one man.

> George Hartzog, on the other hand, believed that he *was* the National Park Service. As one of your sources said, he was a tyrant. He had a vision: a vision which necessarily had to be that of everyone else in the organization. His vision was vast expansion of national parks, but he failed to create a way to sustain them.

When Hartzog died in 2008, the Summer 2008 issue of *Arrowhead*, the National Park Service employees newsletter, featured an article about him on the front page (headlined "NPS Mourns the Passing of Former Director George B. Hartzog, Jr."), which included this tribute: "He was a visionary and his efforts went a long way in enlarging the agency's role in urban recreation, historic preservation, interpretation and environmental education."

21

Saving the Smokies

I received in the mail one day a cordial letter from Sam Vaughan, my editor at Doubleday in New York, asking if I would be interested in writing a book about the Great Smoky Mountains of western North Carolina and Tennessee. Apparently, as I learned later, a Doubleday salesman in the field had determined there was a market for such a book and, after *Whose Woods These Are*, I seemed the logical author for it.

Maybe so, but I allowed time to question whether I was qualified, how long it would take, and whether it was something I really wanted to do. I first saw the Smokies in 1947, soon after working as a reporter for the (Nashville) *Tennessean*. I came to Gatlinburg and found it still off the beaten path, a gentle country crossroads, certainly with little resemblance to the traumatic and tawdry Gatlinburg of later years. At that time I knew little about national parks or southern mountains or mountain people, but when I went to Newfound Gap and saw the bronze plaque marking establishment of the park amidst endless mountain splendor, I felt elevated and enriched in spirit.

I returned two or three years later while working for the American Automobile Association (AAA) on a tour for travel writers hosted by various regional promotion organizations. I was introduced to mountain craftspeople—woodcarvers; weavers; makers of pots, dolls, and dulcimers—and to Cherokee Indians, park rangers, and forest rangers. Tourism was different then, more civic and cultural instead of overblown with commerciality and crowds as it later became.

I remember traveling down the crest of the Blue Ridge, the eastern rampart of the Appalachian Mountains, first on the Skyline Drive in

Shenandoah National Park and then on the Blue Ridge Parkway, which links Shenandoah and the Smokies. I thought there was nothing like it: There were no billboards or high speeds, as on the freeways; I could stop at mountaintop wildflower gardens, restored mills, weathered cabins, and overlooks facing farms and national forests.

Later I was fortunate, in the course of research for *Whose Woods These Are*, to call at byroads and backcountry of Appalachia, entering crannies, coves, and hollows like Gingercake Mountain, Dogback Mountain, Sitting Bear, Hawksbill, Table Rock, and Devils Hole. I saw the region had a unity to it, comprising a cultural and ecological province, yet each and every creek, creek valley, and ridgetop was distinctly its own.

Ultimately I concentrated on the Great Smokies, on the North Carolina–Tennessee border, but the Smokies' composition was but one star in the Appalachian heaven from which it cannot be disembodied. I saw Appalachia as our very own rain forest—no need to look beyond. Everywhere I went in the rich, mesophytic woodlands I found abundant liverworts, mosses, mussels, salamanders, flowering plants, sometimes flowers even in branches of tall trees, forest life varying with elevation, slope, soil, degree of sun and shade.

I recall visiting the Chattahoochee National Forest (which got its name from the Cherokee word for "Flowering Rock," for the many waterfalls tumbling in the highlands), first in spring, with the beauty of the mountains accented by dogwood, redbud, mountain laurel, azalea, and rhododendron. I visited again in fall, a time that combined southern mildness with color changing of the hardwood trees, and with waters flowing swift and clear, cascading down through creeks into streams and out into rivers, their directions determined by the Blue Ridge Divide, carved through the rocks millions of years ago. I saw wonderful wild gardens scarcely known beyond the local communities wherever I went.

I corresponded and talked on the phone with Vaughan. Once we agreed that I would write the Smokies book, he challenged me to raise my sights and foresaw the possible end result as "a small classic," to use his words.

Now many years later, this book, *Strangers in High Places: The Story of the Great Smoky Mountains*, is still in print, having been through several editions. I had no idea it would endure so long, or the meaning it would have for people, or that I would walk into places to be recognized by my name. Once, after delivering a lecture, a man came up to tell me that his mother gave him a copy to take with him on military service in the Pacific, and, when he became homesick, he would open the book, read awhile, and feel better. Another time a fellow said, "I had to come talk to you, as I was sure you was dead."

I think the success of *Strangers in High Places* came not from the coverage of history, geography, geology, or anything purely factual or scientific, but from the portrayal of people living in what was long perceived as a closed mountaineer community. I've been asked how I, as an outsider, was able to make it with the natives. The truth is I can't remember a single unpleasant incident or ever once being turned away. I do remember the pleasures of listening by the hour to the old bear hunter and the mountain musician and the banker in Bryson City, North Carolina, and the mountain doctor and the old Cherokee chief and the shaman and the logger, ranger, and naturalist, professional and otherwise, now regretting only that I failed in those days to travel with a tape recorder. I mean that the voices I heard are gone forever, and that reporting them on paper to be read comes nowhere near as valuable as recording them to be heard.

I did have a problem connecting with the moonshiner in the book. My efforts seemed hopeless, especially after I had been canvassing the backwoods with the revenuer. Luckily, a forest ranger I knew and turned to for help was a connoisseur of mountain whiskey and arranged for me to meet a fellow who made it for a living. During the fateful, backwoods interview that lasted all afternoon, I had to listen to my host say, "Have another," with the only allowable answer being, "It sure is good."

I felt privileged to meet and know fine local people who worked on the staff of Great Smoky Mountains National Park. Superintendents came for a while and then moved on. Maybe the park headquarters could have made it without them, but rangers like Bill Rolen knew the park intimately because they had grown up there, in Rolen's case at Oconaluftee, before it was a park. The same was true of Glen Cardwell. His family had lived in the Greenbrier section that was incorporated into the park; he never strayed far from it. Mark Hannah grew up at Cataloochee; the last time I saw him, he gave me a beautiful poem he had written of what it all meant to him. Arthur Stupka was not a native, but might as well have been. He came to the Smokies in 1935 as the first park naturalist and learned virtually every plant in the park from his field trips.

Of all the people I met, I was influenced most deeply and lastingly by Harvey Broome. He was a Knoxville lawyer, president of the Wilderness Society (one of its founders, in fact), and the leader in the political battle then underway to protect the wilderness of the Great Smoky Mountains. Broome died in 1968 but, many years later, in 2001, when the University of Tennessee Press republished his personal journal, titled *Out Under the Sky of the Great Smokies*, I was privileged to contribute the foreword. "Here we find Harvey, the wilderness apostle, on his home turf," I wrote. "He reveals himself exactly as I knew and loved him: a gentle spirit, sensitive to the needs of nature and humankind, always with tolerance and good humor."

Broome began his lifelong love affair with the Smokies in another age in history. It was already the twentieth century but little changed from the nineteenth. Clusters of rural settlement like Sugarlands and Cades Cove were largely isolated from outside civilization. Mountain people lived as their parents had lived before them. They grazed cattle on the grassy balds, following old pathways. The most significant intrusions into the wild Smokies were then underway by logging outfits and their railroads, but few trails led to the inaccessible peaks.

Following the 1925 Appalachian Trail Conference, individuals and groups along the Trail did wonderful things to advance its goals. The Smoky Mountains Hiking Club assumed responsibility for the length of the Trail across the Smokies, then still wild and little known, at a time when the movement to establish a national park was just getting underway. The club aimed to increase interest in hiking and love of the mountains by disseminating information and taking beginners on hikes, initially scheduled once a month, to key landmarks like Mount LeConte, the grassy balds of Thunderhead and Gregory Bald, the Chimney Tops, and the big trees in Porters Flat. In 1927, the schedule was increased to two hikes a month, throughout the year, and the hiking program enlarged from a leaflet to a substantial handbook. By then, Broome was corresponding secretary and editor of the handbook.

In 1966, Broome lived his finest hours, the summation of a life devoted to the cause of the Great Smoky Mountains and to wilderness everywhere. In that year, the National Park Service announced its first wilderness proposal under terms of the Act of 1964. That law directed federal land management agencies—the Fish and Wildlife Service, Forest Service, and National Park Service—to review potential wilderness under their jurisdiction, to conduct public hearings, and to submit recommendations to Congress for additions to the National Wilderness Preservation System. The agency chose for its precedent the Great Smokies, of all places, but its proposal could not have been worse.

Instead of a plan for wilderness, the Park Service offered a design for roads to solve seasonal traffic jams, including the construction of a new transmountain road that would cross the Appalachian Trail, plus corridors for additional inner loops. What was left over, less than half the park, was offered for inclusion in the National Wilderness Preservation System—in six broken blocks, ranging in size from 110,000 acres down to 5,000 acres. The wilderness proposal was part of a master plan to accommodate ever-increasing numbers with massive campgrounds of two hundred, three hundred, even six hundred units.

Harvey Broome may have been a gentle soul, but he was determined to protect the wilderness and to block the transmountain road. Responding to the challenge of providing for more and more visitors, Broome wrote:

> It must be clear that the demand which now looms over us can never be satisfied. Slow attrition follows development. Almost without exception, wherever there is a road or dug trail or shelter facility in the virgin forest, there is slowly spreading damage. The areas contiguous to developments become littered, eroded, or threadbare from heavy use and abuse.
>
> No further development of any character should take place. No more trails; no more shelters; no more roads; no expansions, extensions, or additions to existing facilities. To protect what is left we must learn to live with facilities we now have. The hardest thing will be the decision itself.

That statement I might expect from an official of the National Park Service, committed to the cause of conservation. However, the antiwilderness design was the personal concoction of George B. Hartzog Jr., the director of the National Park Service, who had made a commitment to local North Carolina politicians for commercial-boosting highways across the Smokies, and who had resisted wilderness designation everywhere. But then, as I observed him, Hartzog was at his best as a political wheeler-dealer and at his weakest as a preservationist of principle. Hartzog evidently thought the Smokies would make an easy beginning, but he could not possibly have anticipated the public's will to be heard. More than two hundred witnesses presented oral statements at hearings in Gatlinburg, Tennessee, and then two days later across the mountains in Bryson City; 5,400 letters were later received for the hearing record. A handful of local politicians and business people supported the Park Service plan, but a parade of preachers and schoolteachers, scholars and scientists, scouts and scout leaders, hikers, trout fishermen, botanists, and birdwatchers spoke for the wilderness. They spoke of the joys of wild places, the spiritual exhilaration, the threats of a political road-building boondoggle. They identified with love of land, idealism, a qualitative experience as the essence of our national parks.

Editorials in newspapers across the country were part of a mobilization of public opinion; it was an education to observe, and an enrichment to feel a part of it.

Broome and his closest comrade-in-arms and old hiking buddy, Ernie Dickerman, gave the spark that fueled the fight. On Sunday, October 23, 1966, a total of 576 national park supporters, members of the Smoky Mountains Hiking Club, members of the Carolina Mountain Club, and other concerned citizens, joined Broome and Dickerman on the "Save-Our-Smokies" hike, walking some portion of the route from Clingmans Dome parking area out along the Appalachian Trail to Buckeye Gap, where the proposed road was intended to cross the crest of the Smokies, and then down to the Elkmont Campground. A total of 234 persons walked the entire seventeen miles, the last completing the trip by moonlight. That hike influenced the decision process. It helped protect the heart of the park from the proposed intrusion.

Dickerman wrote to me years later: "It is amazing how many persons from all over the country supported wilderness designation in the Great Smoky Mountains, and opposed any new roads in the course of the campaign which lasted six years until George Hartzog finally threw in the towel." A report issued in January 1971 declared that the Smokies comprise "a natural treasure of plant and animal life living in an ecological balance that once destroyed can never be restored," and the transmountain road plan was withdrawn.

That made sense to me. I was glad to join in defense of wilderness and to assert that national park policy must transcend local politics and be determined in full public view. I wanted my book to speak for preserving and protecting wilderness, for I believed the Great Smoky Mountains were still one of God's special places—even while much of the world, close at hand and far away, was going downhill—and that the national park became more of a treasure with passing years, deserving the same love and care society bestows on works of art.

Were it not for everyday people who cared, the Smoky Mountains likely would be another parcel of real estate, developed and probably overdeveloped. My interest was in the relationship between mountains and people, how they influenced and cared for each other and how they cared for the earth. Through the research for my book and participating in these activities, I became involved with a new set of mentors who dealt in sheer idealism.

I saw Broome for the last time early in 1968, weeks before he died. A few months earlier he had climbed Mount Katahdin in Maine. Then, near Thanksgiving, he learned that he had a heart ailment. Still, he came to Washington on Wilderness Society business and had dinner with a small group of friends. While walking with him back to his hotel, I saw that he looked pale and weak. The once tireless hiker felt he must stop to rest every few steps. On March 8, 1968, Broome collapsed and died in his yard while sawing a segment of a little hollow log to make into a wren's house.

Over time, I collected and read old and new books, booklets, and periodicals. I visited the Library of Congress and the Smithsonian Institution in Washington, university libraries, plus assorted bookstores and libraries in North Carolina and Tennessee. These excursions opened the door to new ideas and new personalities. I discovered *The Travels of William Bartram*, the record of a thrilling adventure into the natural domain still covering much of the South in colonial days. Bartram was John Muir come alive in an earlier time, and Ralph Waldo Emerson and Henry David Thoreau as well. He was the son of John Bartram of Philadelphia, the plainest of men—a plowman whose curiosity was stirred from observing the interaction of life-forms as he tilled the soil on his farm and who, over time, became a self-taught natural scientist and scholar. John Bartram was, in fact, a principal figure in the botanical discovery of America, appointed in England as botanist to the queen, and a correspondent with Carl Linnaeus in Sweden.

William Bartram shared his father's passion for horticultural adventure and often accompanied him on explorations from Philadelphia to the luxurious wild forest gardens of the South, discovering rare and beautiful plants unknown to connoisseurs and collectors in the North and in England. On one of their journeys along coastal Georgia they discovered a small grove of unusual flowering trees, which John named the *Franklinia alatamaha*, or Franklin tree, for his good friend Benjamin Franklin. William returned later to gather seeds and to propagate the Franklinia in their Philadelphia garden, where its descendants still thrive—though it is no longer found in the wild.

I went to see and be inspired by the Franklin Tree within Bartram's Garden, bordering the Bartram House, which John Bartram built with his own hands in the colonial mid-eighteenth century and which is now part of Fairmount Park on the banks of the Schuylkill River in Philadelphia. William failed in a variety of business undertakings, but he was not meant for them. He found his true calling when he set out to hunt plants and to paint and draw the natural world. Like a precursor of the modern ecologist, William described the interdependence of creatures: the spider preying on a bumblebee, which had lit to feed on the leaf of a plant; a coachwhip snake wreathing itself around the body of a grounded hawk; the battle between crayfish and goldfish; the swarming assembly of alligators feeding upon a vast solid bank of fish. To him, every little thing had purpose, personality, and beauty. He taught friendship even with the rattlesnake, who is "never known to strike until he is first assaulted or fears himself in danger, and then always gives the earliest warning by the rattles at the extremity of his tail."

At a venerable Washington bookstore that featured old and out-of-date government publications, I made two prize acquisitions (which I still own). These were the Fifth Annual Report (of 1883–1884) and the Nineteenth Annual Report (of 1897–1898) of the Bureau of American Ethnology, oversized books of more than five hundred pages each, with maps, photographs, and illustrations of the period. Those works were inspired by the director of the bureau, Major John Wesley Powell, a Civil War veteran and one of the great adventurers in American history. Powell had earlier become the first white man to lead a party in boats down the Colorado River through the Grand Canyon. Then, at the Bureau of Ethnology he endeavored to record and interpret everything possible about America's native peoples before they were lost to civilization.

Powell dispatched one of his most able associates, James D. Mooney, to the Great Smoky Mountains in order to unlock the Cherokee storehouse of prayers, sacred songs, and formulas relating to all of human existence. Mooney collected and labeled more than five hundred plants used by the Indians for their food, medicines, and rituals; I was privy to examine these in their repository in the herbarium of the Smithsonian Institution. Exactly what each plant revealed was beyond me, and yet I felt pleased that it had been saved and I should have the chance to see it.

22

"Wilderness is My Lifestyle"

In time, I found that I was identified, in addition to being a parks person, as an advocate for wilderness and for saving significant portions of it in modern America. It was something I wanted to be and to do. I traveled to areas considered worth saving, and wrote and spoke about them in one venue or another.

I've been asked many times how I got that way, and I've asked it of myself. I did not grow up in rural America with wilderness near at hand, but in New York City, a world of subway trains, skyscrapers, urban beauty, and urban blight. Perhaps I learned to appreciate wildness when I served as a navigator in World War II, plotting the course across oceans by following the stars. I wrote of that experience:

> Night faded slowly. Stars dipped into the dark horizon that lay ahead to the west. Behind the aircraft, to the east, the sun touched a splash of brightness, like an alpenglow, on a new day. Presently the colors of heaven covered a spectrum from pale white through pink and soft blue to the lingering nocturnal black. As a people, we are conditioned from infancy to bathe in light and shun the dark, but one cannot be lonesome, as John Muir once said, when everything is wild and beautiful and busy and steeped with God.
>
> If this is true in the world Muir knew best, of mountains, forests, meadows and glaciers, how much more it applies to the seas. For there is a logical connection between all things in the immense connected body of salty water that covers more than two-thirds of the earth's surface, a grand pattern embracing all the storms and calms, the deeps and shallows, the animals, plants and birds, and the humankind traveling the surface and the skies above and living on the shores of all the oceans.

> As night wore on aboard our little plane and the canopy of the heavens changed, I felt navigation had brought me close to them, that I could sense the harmonious pattern of the universe, of time beyond time and space beyond space joined as one. The celestial navigator can expand the dimensions of his experience, can free himself of the containment, can become part of the continuum of history.

That experience doubtless influenced me when I wrote *Battle for the Wilderness* in 1974. It was published under the auspices of the Wilderness Society, and meant as a primer, or guide, for citizen activists to implement the Wilderness Act. Even earlier, within a year after passage of the Wilderness Act, I went on a week-long horseback survey trip in the High Uinta Mountains in northeastern Utah. I had never heard of the Uintas, yet here I found a rugged, breathtaking expanse of high peaks and flat-top mountains, of brilliantly colored rock formations and of clear rivers plunging into deep canyons. That expedition was initiated by wilderness-oriented fellows I knew in the Forest Service. They hoped the Uintas would become one of the first units added to the National Wilderness Preservation System. But that was not to be, not in the face of conservative, commerce-first Utah politics, at least not until 1984 when Congress designated the High Uintas Wilderness, which now includes more than 456,000 acres.[1]

At one time or other, I met in workshops or seminars with the directors of wilderness and wilderness managers of the four federal agencies mandated to protect wilderness under terms of the 1964 Wilderness Act. These were the Bureau of Land Management, the Fish and Wildlife Service, the Forest Service, and the National Park Service. There were able, committed individuals among them. For instance, I went backpacking in the Idaho wilderness with Tom Kovalicky, a forest supervisor, who, out of the blue one day, made a beautiful statement—simple, yet lofty and profound:

> You get away from your tradition and lifestyle in a wilderness and you find out in a heluva hurry who you are and what you're capable of, what are the real issues in life. What really frightens you will come to the surface.
>
> Wilderness is my lifestyle. Wilderness is necessary. It represents that part of America that once was and always will remain. Wilderness is forever. We should be lucky enough to be

> smart enough to set it aside. We don't have to be like the Europeans. We don't have to wish for that type of land representation. We'll have it. I think we're smart in doing it.

National Park Service people rarely, if ever, talked this way. They may have thought it, but they didn't say it. They didn't dare—not if they wanted to get ahead. Earlier I wrote that, in 1966, the National Park Service chose the Great Smoky Mountains for its very first wilderness proposal, under terms of the 1964 Wilderness Act. However, instead of pressing the cause of wilderness, Park Service director George B. Hartzog Jr. personally pushed for a multimillion-dollar transmountain road across the park, plus additional inner loops and massive campgrounds. He and his agency were motivated not by desire to protect wilderness but to weaken it, so that parks would remain open to mass recreational and commercial development.

The following year, on April 7, 1967, I attended the Sierra Club Biennial Wilderness Conference in San Francisco and heard Hartzog warn that applying the criteria of the Wilderness Act "would jeopardize the whole national park concept." He echoed the sentiments of Forest Service chief Edward P. Cliff, who shared the program with him, about "purity" in classified wilderness: "If we are to preserve the integrity of national park wilderness, we dare not lower its standard or compromise its integrity by the inclusion of areas that express in less than the highest terms the definition of national park wilderness."[2]

23

The Voice Crying in the Wilderness

I continually got on well with Clare Conley, the editor of *Field & Stream*, but then he felt stymied and frustrated by the higher levels and quit. That was beyond me, but I recognized that we were working in the age of conglomeration for the large mainstream media: publishers were swallowed whole by larger fish, only to be devoured again by some mammoth corporate octopus. *Field & Stream* was a typical case, owned by the Columbia Broadcasting System (CBS) and sharing a floor in its Madison Avenue building with Holt, Rinehart and Winston, an old, respected book-publishing firm that had lost its independence and was now another CBS subsidiary.

The new editor, Jack Samson, assured me that life would go on without change, but he didn't mean it, and I was fired in October 1974. It hurt then, and it still hurts all these years later. I felt deeply wounded, but picked up the pieces and moved on.

I decided, however, not to go quietly. Once the word got out, I was flooded with messages, warm and supportive, from former readers, primarily hunters and fishermen. They enriched my life and deepened my commitment. Samson disclosed what he called "the real reason" for my dismissal: that I was anti-hunting. In a widely circulated letter, he wrote, "When someone writes anti-hunting material in another publication and also submits a monthly conservation column to this eighty-year-old hunting and fishing magazine, it is more than sufficient grounds for dropping that column. No one who is anti-hunting will remain on the masthead of *Field & Stream* as long as I am the editor."

As evidence of my presumed anti-hunting bias, he disseminated a photocopy of a page from Cleveland Amory's new book *Man Kind? Our Incredible War on Wildlife*. Amory was a highly successful author and also the eloquent, intellectual founder and president of the anti-hunting organization Fund for Animals. He was the man the gun crowd loved to hate. In his book, Amory quoted from my latest, *Battle for the Wilderness*, in which I wrote that, although hunting plays a valid role as an outdoors experience,

> the need to hunt for food is gone. Much of sport hunting has scant relevance to primitive instincts or old traditions. It does little to instill a conservation conscience. Blasting polar bears from airplanes, hunting the Arabian oryx—or deer—from automobiles, trail bikes or snowmobiles, tracking a quarry with walkie-talkie radios, killing for the sake of killing annihilate the hunt's essential character. There can't be much thrill to "the chase" when there is little chase. At one end of the spectrum, "slob hunters" shoot farmers' livestock, road signs and each other. At the opposite end are the superpredators: jet-set gunners whose greatest goal is to mount on their walls one of everything that walked Noah's plank.

I was simply telling the gun crowd to clean up its act and to eliminate from its ranks those who violate the basic rules of outdoor sportsmanship. Samson's argument, I felt, would convince only fools. *Time*, in its issue of November 4, 1974, devoted almost a full page, headed "A Voice in the Wilderness," with my photograph. The magazine said I was "tough and tendentious," that I had been fired because I made enemies in big business, in the gun lobby, and on Capitol Hill. It quoted Representative Henry Reuss of Wisconsin: "If *Field & Stream* has no place for Frome, then we have come to a time when the voice of conservation is, quite literally, a voice crying in the wilderness."

I put the controversy behind me, but this was not the end of the Jack Samson/*Field & Stream* story. After I left the magazine, Samson set a new tone for it. An editorial he wrote in the January 1975 issue asked the question, "Oil and Water: Can They Mix?" with an answer about cooperating with the great oil companies, keeping their employees working, and avoiding the pitfalls of preservation. It sounded like it came from an Exxon annual report.

Soon after publication of the article in *Time*, I received a call from Frank Sartwell, the editor of *Defenders*, the bimonthly magazine of Defenders of Wildlife. "I am authorized to offer you a column," he said, adding with a laugh, "as long as you don't get us sued for libel."

I already knew *Defenders*. It had carved a place for itself by picking up issues that other groups overlooked. In the early 1960s, when I observed caged bears and snakes under glass at one tourist trap and another, I wanted to find an environmental organization to raise a danger flag and make an issue of it. I visited or called various groups in Washington—the National Audubon Society, the Humane Society, the National Wildlife Federation—but all were too busy with what they felt were bigger things and policy issues. Then, somehow, I discovered Defenders of Wildlife. I had never heard of the organization, but was made welcome. I wrote an article for the January 1964 issue of their magazine, which began:

> Nobody knows how many animals are caged and corralled at roadside tourist attractions which generously advertise themselves with such polite titles as "reptile gardens," "kiddies' backyard," "prairie zoo," and the like. Nobody knows how most of these creatures are captured. Nobody rightly heeds how they are treated in captivity, or fed, how they live, how they die.

That proved only the first step. I learned that Defenders of Wildlife was the one organization opposing the federal government's age-old war on predators—wolves, coyotes, mountain lions, eagles, and other critters that dared interfere with commercial livestock and huntable species. This set me on a new track. In 1969 Coward McCann published my book *The Varmints: Our Unwanted Wildlife*. That book was written for young readers. I remember one day, when I was in the middle of it, the editor called to ask how I was doing. I said I hoped I wasn't writing over the heads of our readers, but he set me straight: "You have only to worry about writing *down* to them."

With Sartwell's phone call, I was a columnist once again. Defenders of Wildlife had come into money from a sizable bequest, enabling Sartwell to edit a presentable magazine with a circulation of about fifty thousand.

With *Defenders*, I was writing for a gentler, more intellectual audience. A writer must find his or her own level, and this was mine. I could ask myself about the moral purpose of writing, my chosen profession, and then answer that it ought to be to express ideals that help society to improve itself. I wrote my column as I wished, through the tenure of three different presidents—all good people who never bothered me. In 1992, when a new president wanted to tell me what to write, I quit.

I wrote many columns about defending wildlife in national parks, and other subjects as well. Titles of my columns included "The Grizzly's Needs Should Come Before Man's" (February 1976); "Yellowstone Lesson: Federal Lands Are Central to Saving Our Wildlife" (October 1982); "Wolves: A Chance to Rectify the Past and Reclaim a Wilderness Heritage" (May 1983); "Instead of Grabbing for Alaskan Parks, Hunters Should Heed a Nobler Tradition" (July 1983); "Are Biologists Afraid to Speak Out?" (July 1984); "Yellowstone Needs Less Man, More Nature" (May 1985); "Neglected Treasure: The Plant World" (May 1986); and "National Park—or Playground?" (March 1987).

To quote from this last column, about Virgin Islands National Park:

> The trouble is that the approved general management plan is a blueprint to keep pace with development rather than to protect the resources at stake. In the section on the Marine Unit, for example, the Park Service identifies visitor activities as boating, snorkeling, scuba diving, windsurfing, sport fishing and boat-camping, and primary users as visitors on chartered yachts and open boats, then stating the planning goal: "Continue the current experience while protecting fragile marine ecosystems and water quality." But these activities are degrading the fragile ecosystem and the water quality. The showpiece of degradation is Trunk Bay Beach, where the well publicized underwater nature trail reaches a depth of only ten feet. Visitors read labels etched on underwater glass plates, but the corals they describe have been destroyed by the sheer weight of too many people.

It wasn't fun for me to write such lines, but at least I didn't get fired for it.

24

Those Endless Compromises Really Do Not Help

I was invited to, and attended, the opening of the Jackson Lake Lodge in Grand Teton National Park in 1955. It was a grand gala event that brought a touch of sophistication to the setting in the wild country of western Wyoming. Assorted local and national luminaries hob-nobbed in good feeling as though all was well, with old issues resolved and deep wounds healed. The cast of characters included Conrad L. Wirth, director of the National Park Service; Laurance Rockefeller, whose family's benevolence had made the event, and the national park itself, possible; and Senator Clifford Hansen of Wyoming, a former governor and former commissioner of the local Teton County.

I was already acquainted with Senator Hansen and we had a good talk sometime during the great day. He was an affable, outgoing, good-humored, conservative politician who, in conversation, made friends, not enemies. Then, especially, he was on his home turf; he was a livestock rancher, proud to have been the "cowboy governor." He smiled and told me he was pleased to be present on this festive day. "I opposed the national park project and fought it for years," he said, "and here I am attending and joining in this celebration."

The federal government had spent millions to put the park together. The Rockefellers spent millions more, including the cost of building the Jackson Lake Lodge, a classy, upscale, modern resort with a bar, a swimming pool, and a huge picture window looking out on a cluster of majestic mountain peaks. The national park clearly put this place on the map as a world-class attraction.

In the days following, I went to see the park from more than the picture window. I was fortunate one day to make a float trip via rubber raft down the Snake River through the heart of the lowland called Jackson Hole. I was doubly fortunate to be the only passenger with the young boatman and guide, Verne Huser, normally an English teacher in Jackson. He was adept at piloting the raft, pointing out the highlights along the way—coyotes, elk, ducks, herons, bald eagles feeding their young in a high nest. He cautioned me to always give wildlife a break by keeping a respectable distance; he felt a responsibility, which he wanted to share.

The Snake was calm, allowing me to shift attention to the huge peaks dominating the landscape. Some mountain ranges are taller, more massive, but the Tetons rise without foothills, boldly and abruptly, parading against the sky, spreading their image across large, forest-bordered lakes. Verne treated me to both geography and history. He explained that the frontispiece, the high basin known as Jackson Hole, was apparently first explored by John Colter, who was a member of the Lewis and Clark expedition and then struck out on his own in 1806, advancing where only Native Americans had passed before. After Colter came mountain men, traders, trappers, and cattlemen, and homesteaders who found the country too high, cold, and barren to farm.

Verne noted that the valley, fifty miles long, was rimmed by mountains—high plateaus of Yellowstone National Park on the north, the Mount Leidy Highlands and Gros Ventre Range on the east and south. It was bisected by the Snake River flowing through Jackson Lake, then through sagebrush meadow and finally through forest and between steep bluffs. From the raft we caught sight of moose, antelope, bison, and mule deer. Huser spotted a pair of trumpeter swans in flight and told me this largest of all swans had been reduced to near extinction, but had made a healthy comeback, thanks to its sanctuary in the park.

We stopped for lunch at a beautiful picnic area on Deadman's Bar, where we exchanged ideas and experiences. We became good friends and correspondents from then till now. I daresay that my choicest acquisition through the years of exploring national parks has been the lasting friendships with all kinds of people. In Huser's case, he went on to run many rivers and write books and articles about them. And we are still in communication, hopefully collaborating for the common good.

In the early 1920s, when the National Park Service was still young and Horace Albright spent his summers as superintendent of Yellowstone, he identified the Tetons as a prime candidate for a new park. When John D. Rockefeller Jr. came to Yellowstone in 1926, Albright brought him to visit this country. Rockefeller shared his vision and wanted to buy the whole basin (as he bought virtually the entire town of Williamsburg, Virginia). Here he established the Snake River Land Company as a front to quietly purchase key tracts. When it became known, the abundant park foes of the area called it a connivance of the federal government and eastern capital. Many locals felt Jackson Hole was meant for growing cattle and hunting, principally by them. They trotted out Wallace Beery, the movie star, to ride across the monument with a rifle in his arms as a gesture of defiance. But the Rockefeller gift ultimately added 32,170 acres of land, presented to the government for preservation.

President Franklin D. Roosevelt, in 1943, proclaimed Grand Teton National Monument from various portions of federal, state, and private lands. The Roosevelt proclamation stirred political opposition from stockmen, loggers, hunters, and the U.S. Forest Service. Wyoming brought legal action to nullify the presidential proclamation. When that failed, Representative (later Senator) Frank Barrett successfully amended every appropriations bill from 1944 to 1948 to forbid any expenditure for administration or protection of the new monument. Finally, in 1950, a compromise bill combined park and monument, transferring certain tracts of the monument to the adjacent national wildlife refuge and providing that Wyoming hunters be deputized in the fall season as rangers.

Many years later I found the country alive with buffalo, elk, deer, and pelicans, as well as signs of grizzly bears. But I found more to it—the hunt as a classic example of the endless compromises made by the Park Service, both willingly and otherwise. The agency had already paid the price in 1950, when Grand Teton became the first national park to allow hunting. Now, each fall, hunters would sign a simple form and become "deputy park rangers."

The rationale was that the elk herd was limited to a winter range less than one-third the size of its range of a century ago and needed to be hunted for its own good. The shooters waited on the flats east of the Snake River for the elk migrating from the summer range in the high country to the winter range in the sagebrush of the adjacent national elk refuge. In an average season of 8,000 to 10,000 elk, they killed about 1,700 and wounded 200, including more than 500 within the park.

Crippled elk, some with legs shot off, were a daily sight. Coyotes, seldom seen in summer, when they pounced on unsuspecting meadow mice, were changed. Hunters saw them feeding on elk carcasses and mistakenly regarded them as competitors rather than as an efficient scavenger patrol.

Rangers who spent months researching and compiling grazing records found certain permits they considered illegal or questionable—the result of inept record keeping, poor administration, and political influence. They found that Clifford Hansen, then a young local county commissioner, had joined with two other cattle ranchers in moving their livestock to the monument lands. Hansen later became a senator who sat on the Interior Committee governing public lands, while his cattle grazed each summer on irrigated public land. They found that while several ranchers held permits, Senator Hansen held the largest, for 569 cattle—the largest permit in the park and in the entire park system.

But the chief ranger ventured to be a political realist. "As long as Cliff Hansen sits on the Senate Interior Committee, we're not going to touch the cows."

Grand Teton National Park is a magnificent natural area. It deserves better than has been given to it. Dams, as a rule, are not tolerated in national parks, yet Jackson Lake is no longer a natural lake, being dammed and controlled by the Bureau of Reclamation to provide water and irrigation to eastern Idaho. The reduction of water flow in winter severely affects aquatic life in the lake and Snake River. Elk hunting is an established annual event, though the Park Service does not allow hunting in any other national park, except for subsistence hunting in Alaska. Fish are stocked artificially. Park visitors fly into Jackson Airport, the only airport within the boundaries of a national park in the lower forty-eight states. And a main stem highway through the park speeds travelers on their way to and from Yellowstone.

25

Fire, Changing Land into Landscape

With two friends in the fall of 1988 I was hiking the trail from Jenny Lake up Cascade Pass in Grand Teton National Park. The monumental wildfires burning in Yellowstone not far north clouded the skies and covered the lake like a shroud of gloom. We smelled smoke and felt it in our clothes.

That was the morning scene. Heading back down the trail in the afternoon, the world turned around. A fair wind blew off the smoke. The peaks of the Tetons shone against a clear blue sky, while Jenny Lake became its old self, reflecting the glories of nature around and above it. The change in a few short hours lifted our spirits. I felt a lesson in the experience, a reassurance that all would be well, in Yellowstone and in our lives, if only we would be patient, allowing nature and time the chance to heal.

Since then, I've thought considerably about the fires of 1988, in a long-term context rather than of the moment. Soon after the fire I met Russell E. Dickenson, a former Park Service director and a good friend, who said, "The Yellowstone we knew for these many years will never be the same." That may be true, but the more I weigh the evidence, the more I look forward to the new, *different* Yellowstone. There is life after fire—fire marks a rebirth, the continuation of a life cycle founded on the harmony of time and patience.

It strikes me that forces like blizzard, cold, drought, earthquake, fire, flood, heat, hurricane, storm, volcano, and wind are both beneficial and inevitable influences on the planet. They shape and reshape land into landscape and reshape form and function of creatures, whether plant or animal, growing on land or in water. They are nature's creative dance, music, poetry, and art that spark creative response in the human spirit.

The damage to Yellowstone was extensive, but more commercial than ecological. "Let it burn," argued the scientists and park preservationists. "Fire will benefit the park in the long run." But commercial, money interests operating inside the park and in nearby communities complained that "fire is bad for business," leading the politicians to demand suppression. The fires became a new source of headline hunting with assorted charges disregarding history in the best interests of the park as a natural life community.

Senator Malcolm Wallop of Wyoming called for dismissal of William Penn Mott, director of the National Park Service, charging that Mott had not been sufficiently aggressive in ordering the fires be fought. Bob Barbee, the superintendent of Yellowstone, was ridiculed by Wallop as "Bob Barbecue." Secretary of the Interior Donald Hodel vowed that if the Republicans were reelected, the government would *really* crack down on national park fires.

In such shallow, self-serving debate, the wonders and welfare of Yellowstone as a particular treasure deserve better—they deserve to be safeguarded rather than exploited for profit and politics. I perceive Yellowstone as the last great stronghold of the grizzly bear, elk, bighorn sheep, and bison, whose interest was ignored in the fire-related hysteria. And so too for the majestic winged animals: the bald eagle, trumpeter swan, raven, and great gray owl, which have long lived with fire and never complained, and which ask our dominant human society to grant them this sanctuary, without complaining.

Yellowstone is more than a parcel of pop culture, more than a collection of geysers, hot springs, and bubbling mud volcanoes serving as tourist attractions, but the embodiment of earth history, complete with upheavals. Millions of years ago the region was submerged beneath an inland sea until the earth thrust crust upward and the water receded. Volcanic explosions and lava flows followed. The earth's crust fractured, great faults developed, and mountain ranges arose. In time, temperatures dropped and glaciers advanced, modifying the hot springs basins, the remnants of vulcanism—a link between the ancient age and our own.

I don't profess to be a scholar in these matters, but it's safe to say that fire has been part of Yellowstone and the West since long before the first explorer came this way. Fire swept through periodically, releasing nutrients and potassium from burned wood. These mixed into the soil to fertilize new growth. Fire opened the cones of lodgepole pine trees, raining seeds on the ground. Yes, most fires killed some trees, but those left, free of competition for nutrients, light, and water, grew stronger and healthier.

Many Indian tribes considered fire a friend rather than an enemy, just as primitive peoples and farmers in various parts of the world to this day set fire "to green up the grass" and stimulate new growth. But as a consequence of wildfire devastation to commercial timber forests, American public policy for the last century has been designed to suppress *all* fires. So the national parks have been managed, although their forests are not intended to produce timber. In 1972, natural regulation, or "let burn," became the policy, allowing natural fires, such as those caused by lightning, to run their course except when threatening human life and property.

That policy worked, reinforced by research showing that most fires die on their own, seldom burning more than one hundred acres. Then came June 1988, when lightning danced in Yellowstone skies. Severe drought had turned trees into tinder; dead litter on the forest floor, untouched during the years of fire suppression, made it worse. The fires took off; a carelessly tossed cigarette butt set off a massive new blaze in July.

In September, when I was in the Tetons, the first snows fell on Wyoming, helping to extinguish the fires. The tourist industry of the region, looking toward its own future, was the first to proclaim that damage, while serious, really wasn't *that* serious. None of the main attractions along the loop road were affected; wildlife was affected only slightly. Plenty of old-growth forest remained untouched; wildflowers and grasses sprung up from nutrient-rich ash.

The Wyoming Travel Commission advised visitors that new national park displays would give visitors a better understanding of the unique role of fire. But the business constituents and politicians need a better understanding than visitors do. I hope we can all agree to treat Yellowstone as more of a national ecological reserve than a national playground. Late in December 2011, I noted a news item of value on this very point. A new report by scientists on the fifteenth anniversary of the return of wolves to Yellowstone showed a quiet but profound rebirth of life and ecosystem health. For the first time in seventy years, the over-browsing of young aspen and willow trees has diminished as elk populations in northern Yellowstone declined and their fear of wolf predation increased. Trees and shrubs have begun recovering along some streams, providing improved habitat for beaver and fish. Birds and bears also have more food. It all adds up for those who are open-minded to fire and wolves, and open-hearted. "Yellowstone increasingly looks like a different place," said William Ripple, a professor in the Department of Forest Ecosystems and Society at Oregon State University, and lead author of the study. "The signs are very encouraging."

26

Yellowstone—Heritage or Honkytonk?

Somewhere, in my travel-writing days, I read a statement by Stephen C. Clark, benefactor of Cooperstown, New York, "the village of museums." "If you offer the people excellence, they will find it out and respond to it," he wrote. That struck home and still remains with me. In due course I discovered that if I consciously strived for excellence, my writing would mature and improve, I would find my subject matter deepen and be more meaningful, and I would come in contact with thoughtful, creative people.

That idea may well have been what brought me to the national parks. I saw in them the best of America and wanted them to be better. In *Changing Times*, the *Kiplinger Magazine*, of November 1962, I wrote a feature titled "America the Beautiful—Heritage or Honkytonk?" I cited the 512-room Canyon Village at Yellowstone National Park, intended as a model in commercially developed lodgings in the entire park system, but I reported that the setting was incongruous with the park, featuring vast expanses of black asphalt, blinking lights over the cocktail lounge, and a gift shop that offered an endless assortment of cheap, imported trinkets that had little or nothing to do with the park.

I wrote on these themes in one magazine and another. I also wanted to encourage Americans to travel wisely, so they would understand and appreciate the parks and bring home worthy memories and mementos.

But this was Yellowstone, which Park Service people refer to as "the mother park," since it was first in America and first anywhere in the world. And the Yellowstone superintendent is second. It deserves better management.

The way that Yellowstone is managed and presented, visitors tend to get wrapped up in it as an entertainment center. The same is true of all the parks, but Yellowstone is largely the model, and that's the way it was laid out in another era. Visitors drive on a designated loop from point A to point B, stopping at one and then another tourist site and convenient concession facility, with scant attention to ecological cost and consequence. This became apparent during an interview I had in the mid-1980s with Roderick A. Hutchinson, the park specialist in thermal geology (the "geyser-gazer," as he called himself). He said that rocks, coins, frying pans, whiskey bottles, engagement rings, sticks, stumps, marbles, shell cases, dentures, bath towels, and clothing, at one time or other, have all been tossed in the thermal features and cleaned out. Curiosity apparently is the principal motivation. Hutchinson summed it up as follows:

> Those visitors don't realize they're damaging something important or that could be important. There are all kinds of little secrets in Yellowstone's thermal basins. As long as we help to preserve them, and allow these small plants and animals to live, something may come along to make our lives better.

27

Concession Power

In his book *The National Parks Compromised: Pork Barrel Politics and America's Treasures*, James Ridenour records a 1990 episode of dealing, as director of the National Park Service, with a powerful park concessionaire.[1] It reveals how things really work and who controls the strings in our national parks. At that time the Yosemite Park and Curry Company was owned by the Music Corporation of America (MCA), a powerful Hollywood movie conglomerate. It may appear like a mismatch, but MCA toyed with Yosemite for a while, even starting a TV series about a park ranger which proved a total flop and was canceled after three episodes. Then MCA decided to dump its park business by selling it to Matsushita, a large Japanese firm. Questions arose about the legality and ethics of selling an American institution to a non-American firm, so MCA hired two legal firms in Washington—one Democratic and the other Republican—to protect and pursue its interests.

The Republican firm of Baker, Worthington, Crossley, Stansberry & Woolf was headed by former Senator Howard Baker of Tennessee, who had been Ronald Reagan's chief of staff at the White House and Republican leader of the Senate. In his Senate days he played a key role in a major environmental issue: successfully circumventing the Endangered Species Act to exempt the Tellico Dam and flood the valley of the Little Tennessee River. That dam has long since proven an eco-tragedy and a colossal failure in all respects. Baker set that behind him; in private practice, his law firm represented a variety of moneyed international clients, including King Hussein of Jordan.

The other firm, Akin, Gump, Strauss, Hauer & Feld, was headed by Robert Strauss, the former chairman of the Democratic National Committee and a high-powered Washington wheeler-dealer. This firm was recorded as employing one thousand lawyers at offices in Washington, New York, and elsewhere in the United States, as well as in Europe and Asia. It proudly asserted that it gave clients not just legal counsel, but "political intelligence."

During the Yosemite negotiations, Ridenour thought he should talk directly with Lew Wasserman, the head of MCA.[2] He telephoned, but the conversation was brief, and a mistake. Soon after, Baker called Secretary of the Interior Manuel Lujan. Ridenour was summoned to the secretary's office, where he found Baker and Strauss present and ready for confrontation. Ridenour recalls:

> They launched into a verbal assault that nearly blistered the varnish on the secretary's desk.... They were angry with me for calling Wasserman directly and not going through them. They were angry that the secretary was publicly saying some things that might screw up the entire MCA deal, reported to be worth $7 billion....
>
> At one point Strauss let us know that they, and they alone, would decide what was to happen to the Park and Curry Company. I vividly remember one remark: "Hell, we will give the company to f—ing Saddam Hussein if we want to!"

That episode shows how concessionaires in the United States' national parks have a distinct outlook that links them with power politics. The basis for it is explained by Joseph L. Sax in *Mountains Without Handrails*:

> At its extreme, in Yosemite Valley or at the South Rim of Grand Canyon, for example, one finds all the artifacts of urban life: traffic jams, long lines waiting in restaurants, supermarkets, taverns, fashionable shops, night life, prepared entertainments, and the unending drove of motors. The recreational vehicle user comes encased in a rolling version of his home, complete with television to amuse himself when the scenery ceases to engage him. The snowmobiler brings speed and power, Detroit transplanted, imposing the city's pace in the remotest back country.
>
> The modern concessionaire, more and more a national recreation conglomerate corporation, has often displaced the local

> innkeeper who adapted to a limited and seasonal business. There are modernized units identical to conventional motels, air conditioning, packaged foods, business conventions, and efforts to bring year-round commercial tourism to places where previously silent languid winters began with the first snowfall.[3]

Yes, our park concessionaires, especially in the large, heavily visited parks, have become very corporate, conglomerate, and powerful. They like to say they are "partners" with park people, but they are definitely the *senior* partners. When they need to, they avoid dealing locally with park superintendents, but go to Washington, and if the issue is really serious they don't fool around with officials of the National Park Service or the Interior department; they choose political connections that count.

The National Park Service administers more than six hundred concession contracts with businesses that provide lodging, food services, gifts and souvenirs, equipment rentals, and transportation to park visitors. These firms pay franchise fees and are represented in Washington by a trade association, the National Park Hospitality Association (NPHA), with James D. Santini, a former Republican congressman from Nevada, as their lobbyist. This is what they believe, as stated in an NPHA brochure:

> Parks should be accessible to everyone: Not everybody can don a backpack and trek across the wilderness. But with responsible management of the parks, and services provided in them, everyone can come and experience the wonder and joy of these "wild" places. From the mysterious enigmas of ancestral Pueblans, to the powerful serenity of national battlefields, to the geologic wonders of Yellowstone's ecosystem, these places are our links to the past as well as windows to our future.
> Visitors' enjoyment is our motivation: Today our parks are more than just a quick vacation stop on the way to somewhere else. . . .
>
> The members of NPHA know that they live and work in special places and take seriously their opportunities (and responsibilities) to share these resources through services to the park visitors.

It grieves me to note that concessionaires do not see national parks as the last representation of primeval life in America. They do not conceive the park as a setting free of human intervention, where the visitor absorbs the "feel" of nature—of plants, animals, and natural features.

They do not see intangible values—of beauty, timelessness, solitude, silence, harmony, awareness—as meant to prevail.

While working as an interpreter in Yosemite National Park during the summer of 1980, my friend Alfred Runte, a well-known historian, talked to visitors about national park ethics and ideology. He began by asking his audience to recognize that national parks are in jeopardy, and would then add:

> What would you be willing to do to see that national parks remain part of the fabric of American society for generations to come? Would you give up some power so that geothermal development would not destroy Old Faithful? Would you use less lighting at home so that strip mines and coal-fired power plants would not be needed in the Southwest?

For his troubles, Runte was directed to a week of "rehabilitation training," if you can imagine that for a scholar. He took it all in and subsequently delivered his message as he chose.

The concessionaires see the park as a business and visitors as customers ready-made for them. Concessionaires own something rarely found elsewhere called a "possessory interest," presumably enabling them to attract private capital and borrow money for construction projects. The federal government owns the buildings in which the concessionaires do business. Thus, if the National Park Service decides a company is performing below standard, it can terminate the contract, but it must buy out its possessory interest. The Park Service also must approve transfers of concession ownership.

This concept has been challenged and fought in the courts. But concessionaires don't lose. The Concessions Policy Act of 1965 assured them the "preferential right" of renewal, and more besides: if a concessionaire chooses to sell, the buyer must purchase his "possessory interest" at current market rate. The Park Service generously accepted the possessory interest concept, believing it would give concessionaires security to invest their own money or to use it for collateral in applying for business loans. In this way, private corporations were given a vested interest in public land—somewhat akin to stockmen holding grazing permits on federal rangelands. The improvements they make enhance proprietary rights and add value to franchises (in the parks) and permits (on the range).

Hotels and camps in the parks are as old as the parks themselves, dating from days when visitors came by stagecoach. Some were built by the railroads or closely connected with them. By the time Stephen T. Mather arrived in Washington in 1915, Yellowstone presented a chain of five hotels and two lunch stations run by one company; three stage lines servicing the hotels; three permanent-camp systems, each offering five camps, two lunch stations and transportation from the northern and western entrances; and assorted traveling camps.

Mather thought they were competitive and chaotic, taking too much parkland. Everywhere, except for Yellowstone, he faced the challenge of attracting capital to the parks. He wanted to improve service and eliminate waste of duplication through a system of supervised or regulated monopoly, protecting the concessionaire as well as the public. The 1917 Yellowstone contract specified the concessionaire's revenue should be "consistent with satisfactory service to the public and a fair return on capital invested, irregularity of seasons, and the otherwise generally hazardous nature of the investment being considered." Concessions became what the Park Service likes to call "supervised monopolies." History shows they are monopolies, but not entirely well supervised.

In the early days, going to the parks was more of a resort vacation or excursion than a wilderness experience for most visitors. In 1877, Yellowstone superintendent Philetus W. Norris established bear-feeding stations near the hotels, using garbage from hotel kitchens to attract bears to entertain visitors. When he was superintendent, Horace Albright built "Greek theaters" of logs at the dumps, patrolled by a mounted ranger armed with a rifle. Those shows lasted until 1948; the garbage dumps finally closed in the 1960s.

Yosemite was the scene of the firefall, where each summer night the concessionaire's voice at the Camp Curry campfire would bellow, "Let the fire fall," whereupon an assistant atop Glacier Point—3,500 feet above and a mile away—would push a huge bonfire over the edge of the cliff. The firefall and drive-through sequoia tree made Yosemite a bit of a carnival, as the bear-feeding did at Yellowstone. But they fit Mather's strategy of image building—of generating public support for his agency.

Some concessionaires in those early years did as much to protect the parks as to promote them. William Gladstone Steel first saw

Crater Lake in Oregon in 1885 and devoted the rest of his life to it. He campaigned to make it a national park, and once that came to pass, in 1902, he operated a concession so he could stay close to the lake. "All the money I have is in the park, and if I had more it would go there too. This is my life's work," Steel said at the 1911 National Parks Conference. Ansel Hall, a pioneer park naturalist, established the concession at Mesa Verde National Park in southwest Colorado, and was succeeded by his son-in-law, William Winkler, a former park ranger who concerned himself with park interpretation until he sold the franchise to ARA, a large lodging and food service company. George Freeman Pollock began his resort, Skyland, in the Blue Ridge Mountains of Virginia in the 1890s, first with a tent camp, then with log cabins. Pollock was the prime mover in the establishment of Shenandoah National Park, by influencing Governor Harry F. Byrd to support the land acquisition with state funds. In 1937 Pollock sold Skyland to the Virginia Sky-Line Company, which largely followed his tradition until it too—like Winkler in Mesa Verde—sold controlling interest to ARA.

Only a few concessionaires concern themselves with the condition of small plants and animals or the little secrets in thermal basins. Most are likely to agree with Rex Maughan in his derogation of elitist ivory-tower advocates and the need to protect their position as regulated monopolies with preferential rights and assured profit. Toward that end they work diligently to sustain proper friends and contacts. And Park Service people are well aware, ever mindful.

During his truncated tenure as director of the National Park Service from July 1977 to May 1980, William J. Whalen tried to clamp down on concessionaires. He charged that in many cases they allowed facilities to deteriorate, failed to provide adequate services, and made excessive profits. Whalen initiated an annual evaluation of concessionaire performance, introduced a new standard contract, and increased the franchise fees. He wanted to end the "possessory interest" through which a concessionaire cannot be removed without a government buy-out. The one to be removed, however, was Whalen.

Other factors were involved, but the concessions contract was critical to Whalen's career, especially after Representative Morris K. Udall, chairman of the House Interior Committee, demanded his dismissal. Udall's intercession was sparked by an indelicate remark made by Whalen at a meeting of concessionaires, but Don Hummel, chairman of the Conference of National Park Concessioners (as it was called then), played a special role in getting rid of Whalen.

Hummel had operated concessions in Glacier, Lassen, Yosemite, and Mount McKinley. He was a former mayor of Tucson, Arizona, and a political supporter of Representative Udall in their common hometown. An attorney and former assistant secretary of housing under President Lyndon B. Johnson, as well as a concessionaire, Hummel was an articulate, combative chairman of the trade association for fourteen years. His viewpoint was always clear. "We've ignored the people. We've protected the animals. We protect the ecosystems, but we haven't put the money where the people can use it," he declared at the 1981 concessions conference. "The Park Service is cutting back the concessioners methodically so that we have less and less security."

Hummel wrote a book, *Stealing the National Parks: The Destruction of Concessions and Park Access,* a rambling personal memoir that he paid to publish in 1987, a year before he died at the age of eighty-one. He wanted to show that parks are not overrun, as charged by environmentalists and their bureaucratic allies in government. According to Hummel, the Park Service reported 346 million total visits for the year 1985, but only 50 million of that total was recorded at the forty-eight national parks. Most of the visits were to urban day-use sites like Golden Gate National Recreation Area in San Francisco and Gateway National Recreation Area in New York; moreover, only 9.2 million visitors stayed overnight in the national parks. Hummel had a valid point, insofar as numbers go. He also insisted that "the private national park concessioners are the true champions of public access and reliable front-line fighters for the right of the people to use and enjoy their parks."

Nevertheless, L. E. "Buddy" Surles, the Park Service chief of concessions management during the late 1970s and early 1980s, was critical of what he considered substandard and unsafe operations by Hummel's concession at Glacier National Park. He felt much the same about the concession at Big Bend National Park in Texas (operated by National Park Concessions, Inc.). Surles moved against them both, but after Whalen's dismissal he got little support. He later said, "It took me two years to get Don Hummel and Garner Hanson [president of National Park Concessions, Inc.] to a point to actually start performing under terms of the contract. I had meeting after meeting with the concessionaires where I went in with a preapproved plan from Russ [Dickenson, National Park Service director] and he'd come in and pour water over the coals and kick my knees right out from me. His philosophy was to give the concessionaires what they wanted to keep his job."

In *Yellowstone: A Wilderness Besieged*, historian Richard A. Bartlett details the concessions saga in the nation's oldest national park. With reference to the longtime principals of the Yellowstone Park Company, Bartlett says, "There is very little evidence that the Child-Nichols interests ever manifested much concern over the park proper. They may be proud of their hotels, buses, boats, and souvenir shops, but to many concessionaires a national park is primarily a place to do business, albeit a pleasant place, and concern is directed to restraints on business or the introduction of a competitor in business, and scarcely ever to environmental threats to the park." Then he commented, "Again and again, incidents appear indicating Park Service cooperation with the concessionaires that seem to place consideration for these businesses above the public welfare."

The history of the Yellowstone Park Company, the principal Yellowstone concession, turned into a comic tragedy. The National Park Service shopped for a new concessionaire and in 1966 proudly presented the Goldfield Corporation, which evolved into the General Host Corporation, an entity distant and uncaring, with profits from a captive audience going out and nothing coming in. Lodgings were poorly maintained and rundown. Roofs, windows, screens, draperies, shades, blinds—all were unclean or unsafe. Employees received low wages and worked long hours. They were housed inadequately, poorly trained and supervised, unhappy in their jobs, and rude or indifferent to guests. Complaints forced the National Park Service to face its own fiasco. A Yellowstone Concessions Study Team in October 1976 conceded, "Visitors are not getting the same quality of service that would be expected or received in most other national parks or in the free enterprise system outside of a park environment."

The government paid off General Host with $19.9 million, then spent almost twice as much to renovate many of the buildings and looked for a new outfit to take over. TW Services became the next concessionaire, followed by the Canteen Corporation, which agreed to invest in upgrading facilities; to operate with a management contract, including a cap on profits; and to allocate a significant percentage to repair and maintenance. Canteen stayed for a while and then was replaced.

Russell E. Dickenson was an old hand who understood the unwritten rule of survival: avoid making powerful enemies or they'll complain and get you. Hummel and Hanson first complained to Dickenson. Then they went to Secretary of the Interior James G. Watt. Thus, at the 1981 concessions conference, Watt assured the concessionaires he was with them, that they would be invited to play a larger role in the administration of national parks. "If a personality is giving you a problem," he told them, "we're going to get rid of the problem or the personality, whichever is faster."

Watt was introduced at the concessions conference with admiration and praise by Rex Maughan, who succeeded Hummel as chairman of the association. Maughan, a self-made magnate, came into the concessions scene through his association with the Del Webb Corporation, which held concessions in Arizona and southern Utah. His principal business, however, was Forever Living Products, a door-to-door network selling shampoo, diet pills, suntan lotion, and skin moisturizers, all based on the aloe vera plant.

Hummel and Maughan shared the viewpoint that parks become meaningful when visited by people; that people must be accommodated and served; and that private, profit-making enterprise is the best way to furnish necessary and appropriate services. In *Stealing the National Parks*, Hummel quoted Mather: "Scenery is a hollow enjoyment to a tourist who sets out in the morning after an indigestible breakfast and a fitful sleep on an impossible bed." Maughan, for his part, stressed the tradition and importance of the "partnership" between concessionaires and the National Park Service.

At a meeting of park superintendents following the 1981 concessions conference, Maughan explained what he meant. He wanted the superintendents to "cut out anything that represents an ivory tower concept." He told them too many decisions were influenced by extraneous, outside-the-park elements, notably environmentalists. "They've tried to convert major areas to wilderness areas, which we feel takes away the majority of the park for most people in favor of providing pristine areas for a minority of park users. Parks are for all the people, not just the environmentalists." Thus, continued Maughan, "if you haven't been to Yellowstone in the winter on a snowmobile, you haven't really seen Yellowstone, and more people should have that opportunity. And if you don't have a profitable concessionaire, you're not going to have a very happy concessionaire. If you don't have a happy concessionaire, he's not going to provide very good services to the public." Surles, a former director of Arkansas state parks, quit the federal government and later took up the ministry.

Many concessionaires feel constantly put-upon, targets of elitist ecologist types working in collusion with park administrators. Consequently they strike back, individually and collectively, locally and in Washington, for what they believe is theirs by right—and indisputably right, as they see it, in the public interest. On the other hand, although commerce is a main pillar of American society, in the national parks commerce colors and clouds meaningful discussion of appropriate human activity and the pressures of increasing visitor use. Private enterprise by its nature promotes business to maximize profit. Entrepreneurs in (and around) the parks generally advocate recreational tourism, from which they benefit, rather than spiritual sanctuary and ecosystem preservation with fitting restrictions and restraints. Wherever visitor service is commercialized, it tends to feature and promote convenience and crowd pleasing, as Sax noted. It leads to advertising to keep accommodations full during the "shoulder seasons," to stay open during winter, to lengthen the "season." It rationalizes selling tawdry souvenirs and package liquor as a way to lower prices for backpacking supplies and peanut butter. The bar trade encourages barroom behavior, brawls, and enforcement problems. Such enterprises and activities influence the entire tenor of a park. It doesn't look like a parcel of primitive America and doesn't feel like one. The park becomes a popcorn playground to which visitors adjust their expectations and behavior. Even when National Park Service employees try to maintain a high level of visitor information and other service responsibilities, that standard is sometimes lost in concessionaire-operated portions of visitor services—with inadequately trained staff, poor maintenance, institutional food, substandard employee housing, and below-minimum wages.

The Park Service is presumed to control concessionaires in terms of service, prices, wages, housing, and maintenance standards. In practice, it doesn't always work that way. Howard Chapman provided significant supporting evidence in an opinion article in the *Marin Independent Journal*, published at San Rafael, California, on January 22, 1990:

> As a National Park Service regional director for seventeen years, I had responsibility for Yosemite during the events that led to the 1980 [general management] plan as well as the years that followed. I have sat in the board rooms of the Music Corporation of America (MCA)—the corporate parent of the Yosemite Park and Curry Company—in the early seventies when we found ourselves outgunned by the financial, political and downright economic aggressiveness of their executives who measured success by the bottom line of each day's profit and loss statement. A national park was to them a place to make money.

Chapman waited to speak until after he retired. Park employees recognize that if a concessionaire wants to implement some new revenue-producing program and meets resistance from the park administration, the concessionaire knows where and how to push the right buttons. Thus, park administrators tread lightly, acquiesce, and rationalize.

Late in 1990, Matsushita, the Japanese electronics giant, responded to a superpatriotic chorus of opposition and backed out of its plans to acquire the Yosemite concessions contract from MCA. The contract was ultimately given to Delaware North Companies, an outfit that outlived a shady past and is identified largely with operations at racetracks, baseball stadiums, and gambling casinos. It employs more than 50,000 people worldwide and does $2 billion in revenue as "a global leader in hospitality and food service with operating companies in the lodging, sporting, airport, gaming and entertainment industries." Little wonder that Sax called the modern park concessionaire "more and more a national recreation conglomerate corporation."

Maughan's Forever Resorts, another case history, owns and operates sixty-five vacation and entertainment properties in or near national parks, recreation areas, and national forests, plus twenty properties in Europe and Africa. It began in 1981 with marinas on Lake Mead and Lake Mohave, expanding to Rocky Mountain, Grand Teton, Isle Royale, Mammoth Cave, Badlands, Big Bend, and Olympic National Parks, plus Blue Ridge Parkway, Padre Island National Seashore, and houseboat rentals at reservoir marinas.

Then there is Xanterra Parks & Resorts, which employs 7,500 people to operate more than 5,000 guest rooms, fifty-five retail stores, and sixty-eight restaurants, as well as marinas, golf courses, and campsites. These include facilities in Yellowstone, Death Valley, Grand Canyon,

Bryce Canyon, Zion, Crater Lake, Petrified Forest, and Rocky Mountain National Parks, as well as Mount Rushmore National Memorial and Ohio State Parks.

There is more to it: the sale of Xanterra to the Anschutz Company, headed by the wealthy Philip F. Anschutz, was completed on September 25, 2008. The interests of this firm are said to embrace real estate, oil and gas development, telecommunications, motion pictures, and sports teams, while Anschutz himself is reportedly a major contributor to conservative causes.

Should Anschutz and Xanterra need legislative or legal advice, plenty is available. That is what Washington is about. They can hire former senators or congressmen, or staff personnel, former secretaries or assistant secretaries of the Interior, or people who make it their business to know people in the right places. This doesn't mean they do bad things, but they know how things get done.

I prefer, and recommend, the route chosen by Horace Kephart, expressed in his book, *Camping and Woodcraft: A Handbook for Vacation Campers and for Travelers in the Wilderness*, first published in 1906 and filled with adventure, personal recollections, and practical guidance. Though in many ways dated, his philosophy and way of life make it enduring:

> Be plain in the woods. In a far way you are emulating those grim heroes of the past who made the white man's trails across this continent. . . . We seek the woods to escape civilization for a time, and all that suggests it. Let us sometimes broil our venison on a sharpened stick, and serve it on a sheet of bark. It tastes better. It gets us closer to nature, and closer to those good old times when every American was considered "a man for a' that" if he proved it in a manful way. . . . It is one of the blessings of wilderness life that it shows us how few things we need in order to be perfectly happy.
>
> Let me not be misunderstood as counseling anybody to "rough it" by sleeping on the bare ground and eating nothing but hardtack and bacon. . . . As 'Nessmuk' says: "We do not go to the woods to rough it; we go to smooth it—we get it rough enough in town. But let us live the simple, natural life in the woods, and leave all frills behind."[4]

28

Charles Eames Had a Word for It

It grieved me to observe many park superintendents believing that concessionaire gift shops were beyond their domain. They tolerated the tawdriest of trinkets on the grounds that souvenir profiteering enabled concessionaires to furnish low-cost lodgings and other less profitable services.

I felt there must be a better way. The superintendent of the Blue Ridge Parkway, my friend Sam P. Weems, had initiated a successful program at Moses H. Cone Memorial Park in North Carolina, introducing craftsmen and women at work with native materials making native products in an appropriate environment. The program gave visitors an enhanced understanding of the people of the region as well as something worthwhile to take away with them. I could point to the same sort of thing at the craft shops of Colonial Williamsburg in Virginia, which produced and sold attractive items of intrinsic value at modest prices. I wanted the shops in national parks to be worthy of their settings—like museum shops—and the Park Service and concessionaires to work with craft groups like the Southern Highland Craft Guild and Indian Arts and Crafts Board.

I wrote a few magazine articles on this theme and from time to time I also took a collection of souvenirs—"the little chamber of horrors collected at America's shrines"—to Washington to show National Park Service Director George B. Hartzog Jr. and Secretary of the Interior Stewart L. Udall, who shuddered. Udall, an Arizona native, tried to help by waiving concession sales fees on American Indian handmade craft items, but that stimulated concessionaires to merchandise carloads of trinkets labeled "Indian," "handmade," and "craft items," which increased their profit and degraded Indian culture. Concessionaires and most park people repeatedly gave the same rationale: they needed something less expensive for children; it wasn't fair to dictate public taste. But it all seemed transparent—the bottom line was maximum profit by cultivating and catering to the lowest common denominator of taste.

Thus, I was surprised in late 1967 when Udall appointed me to an advisory committee of citizens "prominent in their respective fields of endeavor" to review the souvenirs sold by concessionaires in national parks. The secretary asked the committee to recommend a comprehensive program aimed at meeting the reasonable needs of visitors with souvenirs appropriate to the national park environment. The other members of the souvenir committee were Charles Eames, the industrial designer best known for the stylish Eames chair; Lloyd Kiva New, an American Indian textile designer and director of the Institute of American Indian Arts at Santa Fe; and Hilmer Oehlmann, chairman of the Yosemite Park and Curry Company, a reasoned concessionaire who served as committee chairman. Two others, a businessman and a labor official, were listed as committee appointees, but never showed up.

The committee convened first in Washington, then made two field trips: to Grand Teton, Yellowstone, Glacier, and Mount Rushmore, and to Yosemite. I was struck by the caution of park superintendents in any open discussion with concessionaires and by the congenial hospitality provided by the concessionaires. For example, Trevor Povah, president of Hamilton Stores in Yellowstone, wined and dined the committee at his ranch in West Yellowstone, then delivered a severe and stern warning against what he called communistic meddling with the right of free enterprise. That is the way entrepreneurs see things: doing business in a public park is not a privilege but, for them, an absolute right.

Eames and New reached much the same conclusions that I had started with. "Why do you sell this trash to the public that you wouldn't have in your own home?" Eames demanded of John Amerman, president of the Yellowstone Park Company. Amerman sought to brush off the question with a laugh. "Oh well, sometimes you have to sell things or do things you don't believe in." Eames wouldn't have it. He snorted, "I can't imagine Ansel Adams taking a picture he didn't believe in."

The committee chair, Oehlmann, reported to the secretary in a detailed letter dated October 25, 1968. Conferences with park officials and concessionaires, he wrote, were candid and lively exchanges covering standards of quality. The committee made various worthy recommendations. Individual members were free to send supplementary views of their own. Oehlmann felt some of the criticism of concessionaires was unduly harsh and he cited substantial progress by them.

New commented that some concessionaires offered the cheapest kind of souvenir merchandise, not in keeping with the general overall standards of a national public institution; some manufactured-by-Indian products were among the worst. Yes, some concessionaires showed quality products, but often of foreign origin and unrelated to the visitor's park experience. "The banal aspects of the problem," New said, "could best be overcome by park superintendents if they would exercise proper supervision of existing policies having to do with the general mission of conservation and the interpretation of nature havens."

Eames offered three recommendations: (1) keep selection and sale of souvenirs entirely separate from the financial problems of running a food and lodging operation in the national parks; (2) select souvenirs that serve as a reasonable extension of what a national park experience is intended to be; and (3) see that the park superintendents have a conviction of what the park experience should be to a visitor—and the imagination to recognize how that experience can be extended into the selection of a souvenir.

My view then, as now, was that a souvenir must have relevance to the mission of the particular park and of the National Park Service. No concessionaire needs to duplicate machine-made souvenirs and curios that are available immediately outside a park area. I felt the concessionaires and the National Park Service had let down the children who came into the parks in great numbers ready for their first lesson in environmental conservation.

Udall acknowledged the souvenir committee report with appreciation. But Udall departed with the election of Richard Nixon as president and it was all filed and forgotten. I met him later and we discussed it. I believe he meant well, but it was another of those issues that got away. I believe there is still plenty of the old schlock that concessionaires would not have in their own homes. But some improvement has been made through the growing presence and influence of nonprofit natural history associations designed to promote historical, scientific, and educational activities in the parks. I've never seen a single item on their shelves not in keeping with the preservation mission.

Perhaps our committee might have accomplished more, but it certainly brought me in contact and communion with two gifted, creative American spirits. New was an artist and designer who taught and inspired generations of American Indian artists at the Institute of American Indian Arts, an innovative school that he and a colleague founded in Santa Fe. Eames may be best known for designing the iconic Eames chair, with its profound, lasting impact on American design. However, I will remember New and Eames for their grace, goodness, service to a patriotic cause, and companionship.

29

The Outcasts Felt Pain, and Found Salvation

In reviewing my work on this memoir thus far, I ask myself whether I am overly critical. A friend once told me, "A good conservationist can go anywhere and have a perfectly miserable time." Yes, national parks are overcrowded, underfunded, and politicized. Too many visitors litter and wander off the established paths, scarring the landscape for those who come later. Too many automobiles cause congestion and pollution. Off-road vehicles and low-flying airplanes and helicopters pollute the wilderness with noise.

My observations over the years have shown me that all these negatives are true; but the parks are still filled with beauty, grandeur, and all-embracing usefulness, as I have found again and again. That is what makes them worth cherishing and defending.

I pursued this idea thirty years ago while exploring different aspects of national parks in Hawaii. On the Big Island, while climbing to the summit of Mauna Loa, the world's largest volcano, a mountain built by layer upon layer of lava, I thought of the early Hawaiians making their way to the top without shoes, backpacks, or freeze-dried food—perhaps without warm clothing—living close to nature and free of the artifices that clutter our advanced civilization.

Native Hawaiians speak of "Aina," the traditional love of land. Their poetic *oli*, or chants, and the hula recount stories and traditions of humankind woven into the natural universe. The summit of Kilauea is considered sacred, the palace of the goddess Pele. As daughter of Earth Mother and Sky Father, Pele came to Hawaii in flight from her cruel older sister, the goddess of the sea. She found her refuge at last in the volcano, where she has prevailed ever since as goddess of fire. This worldview may be dismissed as superstition or respected as reverence for life—take your choice. But while science may increase knowledge, scientific data cannot be equated with feeling that derives from the soul.

I hiked upward through a forest of ohia, the pioneer tree of fresh lava flows, and heard the rambling, rolling song of the apapane—the most common surviving species of Hawaiian honeycreeper—flitting from one tree to another in the forest canopy to feed on nectar from ohia blossoms. Everything natural and native about Hawaii—its birds, insects, plants, ferns, and trees—and everything so distinctive and luxuriant.

And yet all of it is acutely vulnerable. Of the seventy bird species found nowhere else in the world, twenty-nine are on the threatened or endangered species list and at least three are close to extinction. They survive because they are safeguarded in a national park.

In Honolulu, I went to the USS *Arizona* Memorial, built in Pearl Harbor over the hull of the sunken battleship. I can't imagine anyone, of any nationality (and I toured the site in company with many Japanese), coming away unmoved, without new caring and compassion. The navy established the memorial to protect the final resting place of 1,100 navy men and Marines who lost their lives on December 7, 1941, and are buried with their ship, but the National Park Service now administers the site as a treasure belonging to everyone.

Then I went to the island of Molokai to visit Kalaupapa National Historical Park, where, more than a century ago, a leprosy colony was established to isolate victims who might spread what was then considered a dread disease. You may have read of Kalaupapa in *Father Damien*, the story of a Belgian missionary priest, written by Robert Louis Stevenson. I found the national park, together with three bordering preserves of the Nature Conservancy, a choice, rare fragment of unspoiled Hawaii, protecting rain forests, high cliffs overlooking the sea, deep valleys, and sand dunes, and providing haven for rare birds and plants. At the old settlement of Siloama, I stopped at the Church of the Healing Spring, where a plaque marks a historic event of 1806, when the first residents—the untouchables banished in exile—arrived here. I recorded the wording of the plaque in my notebook:

Thrust out by mankind
These 12 women and 23 men
Crying aloud to God
Their only refuge formed a church
The first, in the desolation that was Kalawao

These powerful words carry a message, not so much of the pain the outcasts felt, but of the salvation they found. And now we all benefit, knowing that Kalaupapa is preserved.

Kalaupapa and Pearl Harbor are special places. They may not cover as much physical space as Yellowstone, but that doesn't diminish their meaning to our generation and generations to come. I believe that national parks like these are meant to be forever; they are priceless time capsules destined for tomorrow that we are privileged to know and enjoy today.

30

A Christmas Gift, Anything but Small

In 1981, after thirty-two years together, my wife, Thelma, and I separated and then divorced. She was a loyal wife and loving mother to our two children. I concede that it was mostly my fault, but we split amicably with little acrimony. I left our home in northern Virginia to spend a year as a fellow at the Pinchot Institute for Conservation Studies at Milford, Pennsylvania, where I completed the updated new edition of my book *The Forest Service*, one of a series on federal departments and agencies. Then I accepted an invitation to join the University of Idaho faculty as a visiting associate professor.

I would be in the Department of Wildland Recreation Management in the College of Forestry. It was the start of a new career and challenge, working with young people in education. My colleagues helped me get off on the right foot by inviting me to deliver the 1982 Wilderness Resource Distinguished Lecture. To quote from my remarks:

> I value the earth for its own sake, not for its utility. The earth is living poetry, music, art, a source of spiritual well being that cannot be found or matched elsewhere. Thus wilderness enables me, in humility, to know myself as part of a limitless cosmos, vaster than our crowded, technological human community.
>
> The artist or poet cannot create a landscape or invent the place. He or she serves only as the interpreter. As Ralph Waldo Emerson wrote, literature, poetry and science are only mankind's homage to unfathomed secrets of nature. I recently read a surprising statement by John Cage, the master of contemporary electronic sound, who said the music he prefers,

> even to his own, or anybody else's, is what we hear if we are just quiet. And what better place to be quiet and listening than the wilderness?

I delivered that speech on December 1, 1982. Christmas was coming. Everybody at the university was busy preparing for the holiday. Everybody, that is, but me. I was alone and lonely. Luckily, my friend Sam West invited me to come spend Christmas at the Grand Canyon. I had met West on a trip to the Grand Canyon. We had serious conversation on the same wavelength and became friends.

West was a red-bearded park ranger and practicing Buddhist. When I went to stay with him, the weather was chilly and snow covered the ground, but skies were clear and bright. We passed our time on the rim of the canyon, hiking partway down, and in West's government quarters, in an old house probably built in Mission 66 days, shared with other park employees. His quarters consisted of kitchen, living room, and bedroom. I had the bed, while he slept on the living room floor. He had some unusual furnishings: paintings and photographs of the Grand Canyon, a beautiful Navajo rug, bronze Buddhist figurines, and a silver and gold tea stand he had brought home from Tibet.

I was there five or six days. Each morning we spent in meditation, sitting cross-legged and quiet, focused on breathing, emptying out the mind. Then West picked up the book of Tao from a shelf and read from it:

> In dwelling, be close to the land; in meditation, go deep in the heart; in dealing with others, be gentle and kind; in speech be true; in ruling be just; in business be competent; in action, watch the timing: no fight, no blame.

"That gave me something I could relate to," West said. "A direct connection with something inside of me. Then I came to the Colorado River. And this place, I think, has a power like no other place on earth. That awakened me further. I became more at home with wilderness, or nature, or whatever you want to call it."

The Colorado River flowing between the walls of the Grand Canyon was West's cathedral. "To stand at the rim of the Grand Canyon, or at the edge of any wild sanctuary, is not the answer. It's so vast, like looking at a picture. You have to go to the inside of it, on a walk down the trail, or a float down the river. Those experiences touch people inside of themselves. The human relationship and interaction are vital."

Yes, they are. Experience has taught me the process is best undertaken alone, in a quiet natural setting free of refinement, a lonely place to search one's soul and restore normal harmony to body and mind.

The idea is contemporary as well as ancient. On August 23, 1981, I was on another trip to the Grand Canyon, commemorating the area as a World Heritage Site, a classification developed by the United Nations Educational, Scientific, and Cultural Organization (UNESCO). A military band played patriotic music and government officials delivered packaged speeches that could have been given anywhere. But I was brought to full attention when the principal speaker, an assistant director general of UNESCO named Abdul Razzak, quoted these lines of James Baldwin:

> For you must say Yes to Life wherever it is found, and it is found in some terrible places. But there it is, and if the father can say Yes Lord, then the child can say the most difficult of words, Amen. For the sea does not cease to grind down rock; generations do not cease to be born and we are responsible to them for we are the only witnesses they have.

The Grand Canyon, for me, was the ideal place to say Yes to Life, particularly at Christmas. But not the only place, nor the only park, as you shall see forthwith.

In 1993, I remarried. My wife, June, and I lived in Bellingham, Washington, about ninety miles north of Seattle, where I was on the faculty of Western Washington University. It was well situated, with the San Juan Islands offshore of northern Puget Sound to the west; cosmopolitan Vancouver an hour north in British Columbia; and the North Cascades National Park, one of our country's last great primeval landscapes, about an hour's drive to the east.

I was well acquainted with parks people, and so each year June and I were invited to the annual staff Christmas party, a festive time with exchange of small gifts. This time, for us, the gift was anything but small. After a ranger won an autographed book of mine, he came over to say his name was John Madden, and he had a special gift for June and me in return—on Christmas Day he would take us eagle watching on a float trip on the Skagit River.

Though we saw but one or two eagles aloft, we were treated to a memorable Christmas; it was so fitting to spend it in a national park, with abundant fresh air and the sound of clean, clear water over the rapids. We were in a house of worship, unroofed, where I could feel anew that I "value the earth for its own sake, not for its utility, for the earth is living poetry, music, art, a source of spiritual well being that cannot be found or matched elsewhere."

31

Muir Found the Icy Wilderness "Unspeakably Pure and Sublime"

Once, at Glacier Bay in Alaska, my three companions—Terry, Bob, and Rick—and I carried with us a copy of John Muir's account, published in *Century Magazine* years after Muir's adventure here. We took turns reading from it as we followed his path, contrasting our times—with our comfort and convenience aboard the *Ice Folly*—and his, recorded as the first white man to enter and explore Glacier Bay in 1879, when he came to Alaska in pursuit of glaciers.

Muir had already challenged the science of his time and proved the influence of glaciation in shaping the features of Yosemite. Now he came near the end of October, when the mountains were mantled with fresh snow from the peaks and ridges of the Fairweather Range down to the level of the sea. Days were growing short, winter was nearing with heavy storms; avalanches would boom down the long slopes and all the land would be buried. When sunshine streamed through luminous fringes of clouds, falling on green waves of the fjord, the crystal bluffs of glaciers and spreading fields of ice, Muir was overcome by what he called "icy wildness unspeakably pure and sublime," asserting that "from all those deadly, crushing, bitter experiences comes this delicate life and beauty, to teach us that what we in our faithless ignorance and fear call destruction is creation."[1]

Muir, like Thoreau, enjoyed storms, considering them exuberant expressions of nature. Still, he chose to travel the main channels and kept warm camping along the forested shores. Muir punished himself, by all standards of normalcy. Who would enter Glacier Bay at the end of October? Because he sought to meet the harsh country on its terms, he was able to evoke lofty images. Revelation derives from effort, as evident in his description of the Fairweather Range: "The white, rayless light of morning, seen when I was alone amid the peaks of the California Sierra, had always seemed to me the most telling of the terrestrial manifestations of God. But here the mountains themselves were made divine, and delivered His glory in terms still more impressive."[2] When he and his canoe party were driven wildly up the fjord, it was as though the storm wind was warning them, "Go, then, if you will, into my icy chamber; but you shall stay in until I am ready to let you out."[3]

He traveled in a huge dugout canoe with three Indians and a missionary named Young. The Indians lost heart with the howling storms. One said Muir "must be a witch to seek knowledge in such a place as this and in such miserable weather."[4] He sternly advised that for ten years he had been wandering alone among mountains and storms, but always in company with good fortune. He said when they traveled with him they need not fear, that the storm soon would cease and the sun would shine, but that only the brave could look for heaven's care.

We followed Muir's course up the bay, passing the Marble Islands, rocky islets teeming with puffins, oyster catchers, guillemots, gulls, and murres. Heading up the Muir Inlet to Muir Glacier, we caught sight of a humpback whale, playfully rolling on the surface, standing on its head, bending in a graceful arc, and plunging below.

My friend Bob Giersdorf had put our little trip together. He was one of my successful friends, who made it as an entrepreneur and marketer of tourism. He always impressed me as having an ingenious touch, devising, designing, improvising, and accomplishing what seemed beyond doing. He operated the concession at Glacier Bay National Park, including the modern lodge, guided fishing, and boating rentals, plus tourist facilities elsewhere in Alaska. Bob combined business success with appreciation of the natural setting. "Our very existence and success," he would say, "depend on making sure that pristine and unimpaired wilderness experiences are preserved for tomorrow, next week, next year and for the next generation of visitors to enjoy. The values here are not only the marine life and scenic wonders, but the quality of experience."

As for the rest of our party, Terry was the director of Alaska state parks, an attractive woman of about forty, a good sport and a good companion. Rick was our boatman, captain, and crew of the *Ice Folly.*

Presently, we hove to near the base of Muir Glacier to watch huge columns and figures of a thousand shapes tumble into the sea, creating heavy swells. Muir described it: "When a large mass sinks from the upper fissured portion of the wall, there is first a keen, piercing crash, then a deep, deliberate, prolonged, thundering roar, which slowly subsides into a low, muttering growl, followed by numerous smaller, grating, clashing sounds from the agitated bergs that dance in the waves about the newcomer as if in welcome; and these again are followed by the swash and roar of the waves that are raised and hurled against the moraines."[5] We were fortunate to see and hear this scene much as Muir had. With global warming in the twenty-first century, much of the ice has melted and gone.

And yet, as Bob mentioned, when Captain George Vancouver sailed through the Icy Strait in 1794, Glacier Bay was only a towering wall of ice more than four thousand feet deep in places, a dent in the shoreline. Muir studied and described the fjords and tributaries resulting from glacial erosion, the imposing array of glacial spires and pyramids.

Finally, we anchored for the night in Blue Mouse Cove. There was little sight of darkness, the sun hanging low over the silent peaks. The next day, after passing Reid Glacier, Rick put us ashore at the edge of Lamplugh Glacier. Terry and I scrambled while climbing over rocks, with Bob, loaded with camera gear, bringing up the rear. At each plateau we rested, listening to the flow of water through crevices in the glacier. Humble pioneer plants covered the rocks. Muir found, at heights, a profusion of flowering plants, with a few grasses and ferns, a demonstration of plant succession.

Thus ends the story of our voyage, but not the story of plant succession at Glacier Bay. On my return home, I read about Dr. William S. Cooper, a young professor in biology and plant ecology at the University of Minnesota, who began his studies of Glacier Bay in 1916. In those days, Glacier Bay seemed as remote as the moon, but Dr. Cooper viewed it as a laboratory of plant recovery in the barren bedrock left by recently retreated glaciers. It revealed a progression from direct chemical extraction from the rock by pioneer plants such as fast-spreading thickets of alder that fixed nitrogen in the accumulating soils, opening the way for more flourishing plants and reforestation.

In Glacier Bay, Dr. Cooper and his students could journey back to the past from mature rain forest to the barren rock at the margin of the glaciers. Once the glaciers yielded, the barrens would become dotted with fireweed, horsetail, mosses, and other pioneer plants, providing the foundation for spruce and hemlock. Then, animals arrived: harbor seals on densely packed icebergs, at the face of Muir Glacier; bears, otters and mink, and mountain goats and hoary marmots in the alpine meadows, feeding on grasses, sedges, shrubs, and lichens.

Cooper attracted scientists to study Glacier Bay in many disciplines. The Ecological Society of America proposed that Glacier Bay be set aside as a natural laboratory for studies of glaciology and biology, and Calvin Coolidge proclaimed the upper bay a national monument in 1925, providing "a unique opportunity for the scientific study of glacial behavior and of resulting movements and development of flora and fauna and of certain valuable relics of ancient interglacial forests."

The work of the early scientists has been built on and expanded by many researchers, resulting in hundreds of scientific publications, including work on glaciology, geology, plant succession, marine ecology, and many studies of marine and land mammals, birds, fish, plants, and fungi. There is nearly continuous documentation of revegetation dating from the 1916 studies conducted by Dr. Cooper and his colleague, Dr. D. B. Lawrence, and their students.

One of those students, Robert Howe, was a friend of mine. He was a ranger and biologist working for the National Park Service on the Blue Ridge Parkway and at Yellowstone, but his dream was to be the superintendent at Glacier, which indeed he achieved and served well.

Glacier Bay, now a national park, covers 3.3 million acres, making it larger than Yellowstone. It is especially popular with cruise ships, floating pleasure palaces. Many passengers enjoy the scenery and the lectures by park interpreters who come aboard. Scientists have conducted studies of tension between whales and humans in cruise ships and tour boats. Whales doubtless would fare better without human presence, but at least the park's vessel management plan moderates the tensions by speed, numbers of ships, noise, and pollution limits. But, I ask, do we need to leave our mark absolutely everywhere?

32

Tourist Boomers Like Action and a Good Show

While reading the August 1987 issue of the *Courier*, the National Park Service house organ, I was struck especially by the article about Voyageurs National Park in Minnesota. Or perhaps I should say I was struck by the headline: "What's Good for Business Has Been Good for the Park." Something there did not ring true.

I could see it reading the other way around: "What's Good for the Park Has Been Good for Business." Certainly that has always been the case. But I believe that when business comes first, trouble for the park must follow, and not far behind—that is, if preservation is the park's purpose. On reading the *Courier*, I felt that preservation might not be the purpose of Voyageurs National Park. Judging from the snowmobile photograph illustrating the article (captioned "Voyageurs as a winter get-away"), one might gather that it served more as a mechanized playground than a pristine park. When business comes first, I can envision the great getaways in store for Glacier, Glacier Bay, the Grand Canyon, the Great Smokies, et al.[1]

That same year, I was invited to speak at a National Park Service conference on tourism in Des Moines, Iowa. It may seem outlandish or bizarre, but the keynote speaker was the president, or executive director, of the National Association of Amusement Parks. He showed little knowledge or appreciation of national parks, but that didn't stop him from telling the parks how to promote more business. The half-dozen superintendents of national park units present at the conference were attentive to him and to his fellow tourist and attraction entrepreneurs. Rarely, if ever, have I seen park superintendents act so friendly with representatives of the Sierra Club or with citizen conservationists.

Years later, in February 2000, I recalled that episode when I addressed the annual meeting of the Voyageurs National Park Association in Minnesota. I was followed by the superintendent of Voyageurs National Park, Barbara West, who spoke from the heart of her desire to leave the park in better natural condition than she had found it. Then she picked up from my remarks, telling of attending a conference of parks and attractions people in Florida similar to the one in Des Moines. She felt uncomfortable and out of place, especially when a spokesman for Disney World reported on the solution to a problem with the display of real-life lions. The problem arose because tourists wanted to see lions move about and behave like bold beasts of the jungle, but the lions were content to lay around at rest all day—until management solved the problem by installing jets of cold or warm air coming up from the ground as needed to energize the lions.

That is the sort of thing tourist boomers would like in our national parks—action and a good show to attract and satisfy the crowds and encourage them to spend. And this beat goes on. Here now, a report in the May 6, 2009, issue of *RV Business*, headlined "National Parks to Benefit from Stimulus Funds":

> The National Park Service (NPS) will use $1 billion in federal stimulus funds to attack an estimated $9.6 billion maintenance backlog among the park system's 388 properties.
>
> "This is going to make a major dent in our backlog. And it will provide jobs," Dean Reeder, NPS national tourism director, told state campground association leaders during the National Association of RV Parks and Campgrounds' (ARVC) 2009 National Issues Conference last week in Washington, D.C.
>
> Reeder said that NPS plans to promote the national parks as a "brand" and establish "a dialogue with the American public" in an effort to stem declining visitation, which has dropped 10% in the last 20 years. He termed this "a crisis."
>
> "It should be a concern to all of us," he said.
>
> Reeder, on the other hand, said that Americans are returning to basics and seeking "more authentic experiences" in their travel. "And that's what we offer," Reeder noted.
>
> As part of the NPS' branding effort, he reported that the park system will conduct a national survey in the fall and establish focus groups to establish the national parks as a brand.
>
> "The fact that people are looking to take shorter trips—in

> both distance and duration—we see as a positive," Reeder said. "The national parks are great and they're doing great . . . except that they haven't kept pace with other alternatives that attract people when they travel. We are getting into that to understand what motivates people and how we can hone our message to reach them at a base level."[2]

It distresses me that national parks should be expected to play such a leading role in the tourist mix of things. It reminds me of an article that I clipped from the *New York Times* travel section of February 22, 1959. It was titled "Invasion of Baja California," and was written by Joseph Wood Krutch, a well-known literary figure of that time. The article was datelined La Paz and began:

> The long, ruggedly beautiful peninsula called Baja (or Lower) California came early into Western history but has stayed pretty persistently out of it ever since. Now, after more than four and a quarter centuries of stubborn resistance to everything called progress, it has begun to undergo what might be cautiously called the beginnings of a tourist boomlet.[3]

Krutch described the "mixed blessings" to "one of the most nearly untouched areas of scenic beauty still left on the American continent." He concluded with a scene in the tropical village of San Bartolo, midway between La Paz and the Cape (at Cabo San Lucas):

> There is one store and an old stone chapel. Surely—although I did not happen to see him—there is also a guitarist. But as I stood by the spring watching the women carrying water, they did not appear to feel put upon. As for the vaquero loitering there on his horse, a good deal more could be said. Held erect in the saddle by the last vestige of Spanish pride, he was certainly not longing for the tourist trade. His son may very well get it whether he wants it or not—but it is not certain he will be better off.[4]

Krutch's theme and style in this article reflected his life experience and maturing viewpoint. In 1952 he quit New York for Arizona and there became a worthy chronicler of the desert and desert life. He wrote extensively about the Grand Canyon and its environs, and served on the board of the Arizona–Sonora Desert Museum. "By contact with the living nature we are reminded of the mysterious, nonmechanical aspects of the living organism," he wrote. "By such contact we begin to get, even in contemplating nature's lowest forms, a sense of the mystery, the independence, the unpredictableness of the living as opposed to the mechanical."[5]

That is what national parks are meant to do: provide contact with the living nature and so remind us of the mysterious, nonmechanical aspects of the living organism.

Early in the twenty-first century I made a personal excursion—following the trail of Krutch from La Paz one hundred miles south to Cabo San Lucas—that celebrated resort at the tip of the Baja peninsula. It certainly wasn't the way the brochures show it. The bus station and unpaved streets in the slums at the edge of town accented the reality of Mexican life. Downtown Cabo San Lucas was a conglomeration of condominiums, construction, and crowds; many Americans in tank tops and tight shorts, and Mexicans hawking their wares and excursions like the Booze Cruise. It conjured the immortal words of songstress Joni Mitchell: "They paved paradise, and put up a parking lot."

That is industrial tourism at work. Natives, like the natives of Baja California, become strangers and servants in their own homeland. In Hawaii, a century ago, residents outnumbered tourists more than two to one; by the year 2000, tourists outnumbered residents by six to one, Native Hawaiians thirty to one. Natives feel alien in a land of mega-resorts with sunbathing, jet skiing, and boozing, where their culture is commercialized and they are expected to work as porters, busboys, housekeepers, bartenders, or pimps.

The National Park Service national tourism director offers a promotion of national parks as a "brand" to keep pace with the public demand for short-order wilderness; never mind Krutch's idea of contemplating nature's sense of mystery, independence, and unpredictability of the living as opposed to the mechanical. That is one way of looking at it. On the other hand, our national parks are undoubtedly the most popular and most loved tourist destinations in America. But, like any object of beauty, a park requires protection, with high standards of care and conservation, to sustain the qualities that make it special.

There I go, preaching again in a memoir. But national parks should never be regarded simply as tourist attractions with dollar signs attached to them. Public recreation is a large and essential factor in contributing to the quality of American life. It serves the economy as well, but that isn't its primary purpose. Outdoor recreation spans a variety of interests, tastes, and goals. Theme parks fill particular niches. So do commercial resorts and campgrounds. But public recreation areas fill a different niche, providing an antidote to urbanized living, a return to pioneer pathways, a chance to exercise the body and mind in harmony with the great outdoors. In such places, Americans learn to understand and to respect the natural environment. Historic parks maintain the opportunity for successive generations to learn firsthand about the conditions that shaped our culture. Contacts of this nature instill the vital sense of being an American.

The national parks assuredly were designed for use by the people, not for an elite aristocracy, nor for scientific study alone. Access to parks is a hallmark of American democracy. But with crowds and jingling cash registers, everything changes. Beauty spots in this country and other parts of the world have been overexploited and milked dry long before their time. Golden arches, chain motels, convention centers, highway strips overcommercialized with billboard blight, and honkytonk tourist traps—they are lookalikes that blot out distinctiveness and beauty.

Fifty years ago, Gatlinburg, Tennessee, was a friendly country crossroads and gateway to Great Smoky Mountains National Park. Today, the "number one mountain resort of the nation" is like an obstacle course of money machines called "family fun attractions." Elsewhere in the country the most cherished national battlefield parks—Gettysburg, Antietam, Manassas, Fredericksburg, Chattanooga, and Vicksburg—have all been tarnished by commercial attractions, saloons, souvenir stands, subdivisions, and condominiums encroaching from surrounding lands and sprinkled throughout private inholdings within the parks.

Each park needs to preserve the fundamental values of its natural ecosystem and historical integrity; otherwise, the inroads of cumulative damage are inevitable. Inch by inch, losses are accepted because they seem inconsequential, but, as at Gatlinburg, Gettysburg, and a hundred other places, they add up to a loss of values that can never be replaced.

The point is that national parks cannot be all things and still be national parks. Prudent and intelligent people must realize that unrestrained pressure on the parks for profit is not progress. It serves to make one generation rich and to impoverish the future. A place of beauty is like a theater; it may be built to seat five hundred persons—if it is, you don't try to cram a thousand persons into those five hundred seats, or give them free rein to do whatever they want.[6]

It's about business, boosterism, getting more people to come and spend more money, not about why the management plan aims to close off and conserve. Yet the emphasis needs to be on protecting and enhancing the quality and character of each park, and letting dollar values follow. When the desires of business interests for profit are allowed to dominate, the beauty will be lost—inevitably and without fail.

In the previous chapter, I quoted my friend Bob Giersdorf, who operated the concession at Glacier Bay National Park in Alaska: "Our very existence and success depend on making sure that pristine and unimpaired wilderness experiences are preserved for tomorrow, next week, next year, and for the next generation of visitors to enjoy." That makes for a sound approach. It preserves the parks and helps to protect a valuable economic resource, rather than squandering and ruining it.

33

The Scientist Who Speaks from Conscience Pays a Price

On the whole, concern for long-term preservation in our national parks comes more from staff personnel—that is, archeologists, biologists, ecologists, architects, and historians—than from administrators, or managers, who are involved with immediate public and political issues. Administrators want the parks to be popular, loved, and supported, always with more visitors, whereas scientists want them to be pure.

Adolph Murie as a case history will help me to illustrate this point. Starting in 1922, Murie spent thirty-two years working for the National Park Service, including twenty-five summers in field research at Mount McKinley (before it was renamed Denali). From April to October 1939, he walked 1,700 miles in his field study of the relationship between wolves and Dall sheep.[1]

He returned the following year for fifteen months in the field, traveling on skis and snowshoes in winter. He knew the birds, wolves, coyotes, lynx, foxes, caribou, moose, sheep, and other species from intimate experience with them in the days before research was conducted by harnessing the animals with electronic devices.

Murie learned by living among the animals, on their own ground. Once he wanted to test the temperament of wolves. He approached a den, reached in, and succeeded in picking up and withdrawing a pup, for the wolves were friendly and trusting. After the mid-1950s he concentrated on grizzly bears, often following a family for days so that he knew and understood each member. He wrote monographs, handbooks, and natural history articles for both popular and scientific journals.

He heeded the advice of his older half-brother, the noted biologist Olaus Murie: "It seems to me we should get away from the strictly scientific method of today, so much like the laboratory technique. We have to speak the truth but we can use human language in doing so." Adolph concurred, "I have, I think, avoided the ecologist's jargon, the scientific phrases so frequently created by ecologists and animal behaviorists to make simple facts sound profound and impressive."

The Murie brothers thought much alike.[2] They believed in naturalness in their lives and their work. Olaus wrote, "Poisoning and trapping of so-called predators and killing rodents, and the related insecticide and herbicide programs, are evidence of human immaturity. The use of the term 'vermin' as applied to so many wild creatures is a thoughtless criticism of nature's arrangement of producing varied life on this planet."

And Adolph, in the foreword to *The Mammals of Mount McKinley*: "No species of plant or animal is favored above the rest, and they grow together, quietly competing, or living in adjusted composure. Our task is to perpetuate this freedom and purity of nature, this ebb and flow of life—first by ensuring ample park boundaries so that the region is large enough to maintain the natural relationships, and secondly to hold man's intrusions to the minimum."[3]

His superiors gave Adolph little support or encouragement. On November 8, 1956, he sent a fourteen-page memorandum to the superintendent of Mount McKinley National Park commenting on the plans for Mission 66, the ten-year national park system development plan. He urged open discussion, with guidance and assistance from conservationists and others outside the Park Service. He was concerned about the future of wilderness and felt the construction of lodgings and the intrusion of a road across the park, with everything on both sides of it labeled like a museum, were grave threats to the area.

For his troubles, he was brushed off by the park superintendent of the time, Duane D. Jacobs, with this response: "I think it is quite reasonable for anyone of your many years of intimate knowledge of McKinley purely as a wilderness area to be somewhat alarmed as Mt. McKinley finally emerges across the threshold of a new era, that of a great national park set aside for the use and enjoyment of the people, which is soon to receive this intended use and enjoyment."[4]

Other scientists have received essentially the same response. In 1973 John Craighead reported that he and his twin brother, Frank, had recorded ninety-one grizzly bear deaths within the Yellowstone ecosystem during the two preceding years, more than double the birth rate, leading to the conclusion that the population was in very serious trouble.[5]

If respected authorities like the Craighead team raise some significant doubt or challenge, then the administrator is well advised to follow a cautious, conservative approach. But in Yellowstone in 1975, the park superintendent, Jack Anderson, brushed them off in the same manner as Superintendent Jacobs in dealing with Adolph Murie.

Providing sanctuary for wild creatures surely ought to be the single most important role of the parks at a time when diversity on the planet is so thoroughly endangered. The National Park Service should be the apostle of wildlife preservation and the advocate of a pesticide-free landscape. But this has not been the case. Until the mid-1930s most of the so-called predators in Yellowstone—principally coyotes but also mountain lions, wolves, and wolverines—were trapped, poisoned, or shot on sight. Ultimately, coyote control was stopped, over the objections of the park staff, who insisted that those villainous critters disrupted the naturalness of the good animals—deer, elk, antelope, and bighorn sheep—on which they preyed.

The view of park administrators reflected outmoded federal programs that emerged from the bounty system—a subsidized public service benefiting the livestock industry—and were biologically unsound, costly, and corruptible. Although wolves had inhabited the entire western half of the United States, by the mid-1930s, under the bounty system, they had all but disappeared.

The tide began to turn with passage of the Endangered Species Act of 1973. To their credit, park personnel ultimately stood up to explain that predators take only small numbers of the animals they prey upon and probably are beneficial to a healthy population. They asserted predation as part of the harmonies of life, as the Muries had taught. Thus, in 1995, despite considerable local opposition, Yellowstone released a total of fourteen gray wolves (while others were released in central Idaho), opening the way for wolf recovery. By 2008 an estimated 1,500 wolves were present in Montana, Idaho, and Wyoming.

Adolph Murie pleaded for purity in the parks because he believed them to be imbued with special spirits deserving human acknowledgment and protection. In an article in *National Parks Magazine* of July 1965, he decried improvement of the ninety-mile road through the heart of Mount McKinley National Park. He saw the old road as being in harmony with the enjoyment of flowers, lichens, wandering tattlers, and grizzlies, charming all visitors seeking the sublime, while the new road clearly was designed to dominate, creating a conspicuous scar over many miles of landscape and showing an obsessive regard for superhighway standards and a lack of appreciation for the spirit of wilderness.

Again, Murie received scant encouragement. His strongest support came from Charlie Ott, who worked for years as a maintenance man in Denali until his retirement in 1970, all the while taking magnificent pictures of wildlife that have illustrated books, won awards, and ranked him high among photographers of animals in the wild. In 1986 when I was in Denali visiting him, Ott recalled that, in his early years, wildlife was easy to see in large numbers and diverse species. He estimated that less than a quarter of the wildlife remained in the park as compared with 1950s numbers; as visitor numbers rose, and development with them, both inside and outside the park, wildlife declined.[6]

Ultimately, the Park Service instituted restrictions on the park road and installed a system of shuttle buses that allows people to get on and off along the way. This has helped to some degree, but questions remain regarding the influences of human activity on wildlife. For instance, what are the behavioral effects when solitary species like bear and wolverine intrude into the territory of their neighbors? Tourism the world over has affected the habitat of wild animals and modified their behavior. The road, the bus, and the fumes; the sight, smell, and sounds of humans, and harassment by amateur (and professional) photographers—these unnatural influences are not understood, yet little if any attention has been directed toward comprehending their effects.

In 1980 Denali was enlarged from 1.9 million acres to 6 million acres, as Adolph Murie had long urged. However, pressures have increased many times over and have led to drastic changes. First came the modern ninety-mile road through the park, then the main-stem highway linking Denali with Anchorage and Fairbanks, Alaska's major population centers. Within thirty years the yearly number of visitors rose from less than one thousand to more than four hundred thousand.

Many arrive by train or bus on package tours, overnighting at standardized tourist hotels at the park border and spending a few hours sightseeing in the park as their brief experience in wild Alaska before continuing on their itinerary.

For most visitors, disturbances such as highways and snowmobiles are acceptable because they generally add to convenience; besides, their negative effects on wild nature are scarcely understood. Nevertheless, as George Wright, the pioneer national park biologist, and his associates noted in 1933, the natural heritage truly comprises more than surface scenic features, but "the intimate details of living things, the plants, the animals that live on them, and the animals that live on those animals."[7]

That view has a long way to fully take hold. Although grizzly bears, as a case in point, depend for survival on national parks (and wilderness areas in national forests), park people think their job is to protect visitors from grizzly bears. In 1975 the Fish and Wildlife Service classified the grizzly as "a threatened species." But it also called *Ursus horribilis* "an aggressive animal that is highly intolerant of man." That kind of prima facie assumption overlooks, or ignores, Homo sapiens as the most "horribilis" of them all, an aggressive critter intolerant of anything that gets in its way. A bear rarely attacks unless wounded, provoked, or startled, or unless it thinks its home, food supply, or family is in danger. When bears aren't bothered, they are less dangerous than most animals; in fact, they generally prefer to avoid contact with humans. Of course, bears are never to be fully trusted because they are wild, but they certainly ought to be respected on their own terrain. Instead of conditioning bears to be respectful of humans, condition the humans to respect bears and bear country. The presence of grizzly bears is a way of nature signaling that natural law remains in force, at least in the national parks, and that these chosen fragments of the planet have not been fouled by the disorder of contemporary human society.

The least management of wild nature is the best management. "Wildlife managers want to manage everything," Adolph Murie wrote in *The Grizzlies of Mount McKinley*, "just as a forester wants to practice forestry in parks, and engineers want to build more and more wide roads."[8]

The scientist who speaks independently from his conscience pays a price. In 1958 Adolph Murie was directed by his superiors to move to

Medford, Oregon, to spend his winters. He did not want to relocate from Jackson, Wyoming, where he had all his books and papers, but he was warned that he would not be able to continue his Alaska studies unless he accepted the transfer.[9] As independent research biologists, John and Frank Craighead, between 1959 and 1971, studied the movement of bears in Yellowstone through the use of radio-tracking collars, a system they pioneered. When their findings differed from official policy, they were subject to harsh criticism and denied their permit for research in the park.

In the early years, the National Park Service published the Fauna Series, including *Ecology of the Coyote in Yellowstone* and *The Wolves of Mount McKinley*, booklets designed to interpret field studies in popular language that anyone could understand. But times changed. The National Park Service continues to employ a corps of scientists and resource management specialists who gather baseline data and discuss ecosystem values. Sometimes they are heeded; sometimes they are overridden by political or bureaucratic considerations.

In 1966, Adolph Murie felt that his responsibility as a scientist transcended his long employment by the National Park Service. He openly protested spraying in Grand Teton National Park designed to eradicate bark beetles preying on lodgepole pines. Park foresters had promoted beetle control for years, but Murie disagreed, citing the so-called epidemic as the high point in beetle population in the historic beetle-lodgepole cycle. He reasoned that the beetle serves as a thinning agent to make room for more lodgepoles or for firs and spruce, and when beetles run out of old, susceptible lodgepoles, the cycle subsides naturally. In the meantime, dead and dying trees provide favorable perches for hawks and owls and habitat for woodpeckers, nuthatches, and brown creepers. Moreover, with opening of the woods, vegetation takes on more variety to become a genuine rather than a phony landscape.

Some park officials couldn't see it that way. One explained that spraying was meant to save the natural scenery. "Will it be natural after spraying?" asked Adolph. The reply was, "Well, the brown needles of dead trees might cause visitors to think the Park Service negligent in its custodianship."[10] That viewpoint, and the extensive use of pesticides, is still widespread. Some park officials feel compelled to provide a sanitized environment that looks natural but is nonetheless free of poisonous plants, pesky bugs, and dangerous animals.

Douglas Larson provides another case history. He began his studies of the water in Crater Lake National Park, in southern Oregon, as a graduate student at Oregon State University in the late 1960s. He was intrigued and fascinated by the brilliant blue of the calm, mirror-like water. Here, in fact, was the deepest lake in the country: 1,900 feet deep, formed in an immense crater twenty-three square miles in area, where an ancient volcanic mountain peak once stood and then collapsed. The deep-blue color resulted from the penetration of sunlight to depths of three hundred to five hundred feet and the "backscattering" of this light by short wavelengths. That certainly made it special and well worth the study.

Crater Lake is not among the most renowned or most visited of national parks, but it is one of the oldest, established in 1902 through the efforts of William Gladstone Steel, who labored for years building support and scientific data. In contrast to other great American lakes, like Tahoe, that fall in the category of paradise lost, Crater Lake, thanks to Steel and other devotees, retains its natural character.

Like Steel, Larson became a Crater Lake champion. His story is well documented in *Crater Lake National Park: A History*, by Rick Harmon, a former editor of the *Oregon Historical Quarterly*.[11] It shows that he collaborated with the park administration and endeavored to provide scientific research about the lake, but that in time turned from cooperator to a severe critic.

From the outset in the 1960s, Larson worried that the lake's exceptional properties might eventually be degraded by the impact of many thousands of visitors at the Rim Village. In the late 1970s and 1980s, he voluntarily launched the park's first serious, sustained effort to study the lake. His research focused on the lake's more than one hundred species of microscopic phytoplankton, plus a program to chronicle changes in the lake's clarity over time.

When he found a significant decrease in clarity, he and other scientists asked whether the cause of it came from an increase in plant nutrients related to sewage leaching from the park's septic tank drain-field system. Park officials, however, offered little encouragement or support. "They didn't want to hear any of this. It sounded bad," noted John Salinas, a park interpreter who worked with Larson.[12]

The sewage disposal issue that Larson raised came to the fore on its own in 1975 when clogged sewer lines near Crater Lake Lodge caused raw sewage to overflow and mix with snowmelt running into the park's main source of water. Park residents and employees became ill from contaminated water. The park was closed for three weeks, the water system was flushed and sanitized, and the lodge remained closed for the rest of the season.

Then, wrote Rick Harmon in his history of Crater Lake:

> In the fall of 1987, an editorial and article published in the *Eugene Register-Guard*—both showing the imprints of Larson's arguments and concerns—brought the issue into broad public view. Far less conspicuous, but just as significant, a Park Service report in the fall of 1987, based on the findings of a panel of scientists, conceded that 'elevated nitrate concentrations' in springs entering the lake were probably caused by the septic-tank leach field in the Rim Village area. The report recommended shutdown and replacement of the sewerage system. However, as the *Register-Guard* reporter observed, superintendent [Robert E.] Benton was "defensive on the [water] clarity subject and castigates reporters for taking such an interest in it."[13]

Harmon wrote that Larson was "increasingly marginalized (if not vilified) by the Park Service."[14] For example:

> Invited to give a paper at an American Association for the Advancement of Science symposium on Crater Lake at Oregon State University in June, 1988, Larson presented "Limnological Response of Crater Lake to Possible Long-Term Sewage Influx," coauthored with biologist Clifford Dahm and botanist Stan Geiger. After the meeting, as the AAAS was preparing to publish the symposium's papers in a volume entitled *Crater Lake: An Ecosystem Study*, superintendent Benton and other park service officials pressured the book's editors to exclude the Larson paper. Larson explained: "Fortunately, the senior editor, Ellen Drake, who's an oceanographer at OSU, publicly stated there would be no censorship of this book."[15]

Larson continued to make his point through the years. As recently as March 17, 2009, in a commentary in the *Oregonian*, he wrote:

> For 30 years and perhaps longer, Congress and successive administrations have treated the national parks not with dignity, but with an attitude that borders on contempt. Park Service budgets were slashed to levels that made it increasingly difficult to even keep the parks open. Buildings, roads and other

> infrastructure deteriorated for lack of maintenance. Personnel ranks were reduced to skeleton crews. Contempt was obvious during the recent [George W.] Bush administration, when efforts were made to (1) open up Park Service lands to developers, extractive industries and other commercial and recreational interests; (2) weaken environmental rules and the Park Service's conservation mission; (3) suppress scientific information; and (4) outsource Park Service employment to private contractors.
>
> Scientific research and essential programs to monitor the environmental impacts of human encroachment were curtailed. And Park Service ideals—protection, preservation, professionalism—took a back seat to mere survival. . . . The National Park Service, like any bureaucracy, also has its share of incompetence, dishonesty and stonewalling.[16]

But he continued through one channel or another to urge a better scholarly approach to the effect of sewage on water clarity: "Crater Lake National Park should have a lake-monitoring research program that is base-funded every year. It should be done routinely . . . future investigators will benefit from having a long-term and thorough historical record with which to compare their data."[17]

In a personal communication to me on October 12, 2009, Larson summed it up:

> During my nearly ten-year campaign to force the Park Service to address the sewage issue, I was shocked by the agency's efforts to not only silence me, but destroy my professional reputation and credibility. . . . As the sewage issue grew more heated, my status at Crater Lake soon became that of a pariah. I found myself isolated and condemned.

In this same vein, I submit here a 1969 study by Edward C. Clebsch, professor of botany at the University of Tennessee, "concerning the values to science of wilderness." Clebsch prepared his paper as part of the effort of citizens to prevent construction of a highway across the Great Smoky Mountains as proposed by the National Park Service.

Clebsch gave reasons why the road should not be built. It would: split a wilderness in two parts; alter the chemical constituents of several fine streams; cause physical changes in the streams by siltation, with only gross consequences predictable; alter ground-water patterns for an unpredictable distance from the cuts and fills necessary to build the road; require that cuts and fills be stabilized by plants, most likely exotic to the region (as are many plants commonly used for such purposes); create a barrier to the movement of animal species; provide avenues for the introduction of new pests and pathogens into the adjacent wilderness; and serve as a point source for contamination and disturbance of the adjacent wilderness by people and their refuse.

Neither a single national park scientist nor resource specialist was heard from when the highway plan was announced. It remained for Clebsch to come forward and declare: "I regard it my responsibility to show the ecological consequences of poor stewardship of one small fragment of our already small wilderness parkscape. . . . The benefits of the proposed transmountain highway to the community of scientists is virtually nonexistent. I believe the liabilities of such a road outweigh the benefits—not just to the scientists, but to the population at large as well."[18]

The strongest point Clebsch made was that eminent world scientists, even then in 1969, were alarmed at the rate of destruction of plant cover and foresaw serious problems and global consequences. Thus, while the amount of green cover to be destroyed by the transmountain road might be very small, when added to the cumulative destruction in other places, its significance would grow.

No such data or warning was provided to the public by National Park Service scientists. Not when their director in Washington was aggressively promoting the Smoky Mountains road, and when political convenience and expediency often govern decisions. Plainly the scientist who insists on speaking independently from his conscience pays a price. There should be a better way.

To summarize, Steve Gndiak, a veteran national park biologist, in 2009, after more than twenty years of service, felt utterly frustrated and chose to retire with this letter to the Glacier National Park superintendent:

> How can we seriously argue that we are protecting wildlife when we can't even answer simple questions about their status? For nearly every species we lack the most basic population information on which to base an inference about declines; climate change or other threats that could sweep whole suites of species into oblivion before we have any chance of mustering a response. How is that serving the public?
>
> If we save the roads and hotels but lose a few species of wildlife, or whole ecosystems, will we have lived up to our promise to protect the parks for future generations?

34

The Vail Call to Arms, Unheard

In October 1991 I spoke at the celebration of the seventy-fifth anniversary of the National Park Service held at Vail, Colorado. That conference was a major event in the agency's history, well attended by Park Service personnel and by prominent citizen conservationists. It was meant to plan "National Parks for the 21st Century" and did indeed produce a 137-page document that became known as the Vail Agenda. James Ridenour, then the National Park Service director, termed it "a great call to arms," since it reaffirmed resource stewardship and protection, public access and enjoyment, education, science and research, professionalism in management, all those good things. But if you ask me, the Vail Agenda is still waiting to be heard and heeded.

Readers who have followed my path may wonder how and why I was invited to speak on the program. Certainly, for a while, I wondered too. I believe the initiative came from three members of the conference committee. One was Paul Pritchard, president of the National Parks Conservation Association, a friend who was always giving me a platform to say things that perhaps he could not feel free to say himself. Another was Nathaniel P. Reed, a Florida activist and a former assistant secretary of the Interior, who introduced me when I spoke. He wrote to me in 2010, recalling his period in Washington:

> Walter Hickel, Secretary of the Interior, was fired after leaking a letter indicating his concern for continuing the war in Viet Nam [*sic*]. After Hickel's departure, for reasons never clarified, Frederick Malek, the president's "Chief Enforcer," took over the secretary's working office and called in a pre-prepared

> list of presidential appointees and fired them all for no apparent reason. Perhaps it was to show the world that the Nixon administration brooked no opposition to any of its policies. Dr. Leslie Glasgow, a distinguished professor from Louisiana State University, had been Assistant Secretary of Fish, Wildlife and National Parks. A gentle, scholarly man, he was among those summarily dismissed.[1]

That is the way the business of government is done in Washington, like a game played every day. Reed himself was serious, committed, and conscientious. While still early in office, he promised me he would give attention to a wayward agency called the Division of Wildlife Service, which, as a branch of the Fish and Wildlife Service, came under his jurisdiction. It was an old, entrenched outfit, a band of "government trappers," or "gopher chokers," paid by the federal government to wage an unrelenting, widespread war against predators, or "varmints," including coyotes, wolves, and bears. The trappers had been working at it for years, mostly for the benefit of the western sheep industry, spreading toxic poisons with scant reference to ecology, wildlife science, or modern times.

It took Reed time, but he made good on his pledge to eliminate the gopher chokers. Or so he thought. The following year the agency was totally reconstituted and relocated in the Department of Agriculture, where it has continued, ever since, to provide a subsidized service to an industry. That is how things work.

As president, George H. W. Bush showed little interest in conservation or the Department of the Interior. Manuel Lujan Jr. had given up his seat in Congress after serving twenty years and was ready to return to New Mexico when Bush appointed him secretary. Lujan came on as a folksy conservative, affable, nonconfrontational, and woefully ill-informed about basic resource questions. He became a handy target for criticism and ridicule; the media referred to him as "the Inferior Secretary," "gaffe-prone," "the Cabinet's weak link," and "one of Bush's dimmer points of light."[2]

Clearly Lujan was appointed to take orders, not to give them, while holdovers from the James G. Watt and Donald P. Hodel era at Interior pressed on with the Reagan agenda, including commercialization of national parks. Lujan apparently had little or nothing to do with the selection of James M. Ridenour as director of the National Park

Service. Though he had never visited the Grand Canyon, Yellowstone, or Yosemite, Ridenour was on a first-name basis with Vice President Dan Quayle, for whom he had raised funds as a county finance chairman of the Indiana Republican Party.[3] Ridenour lacked background, but he had graduated from the University of Indiana with a major in recreation and had worked as director of natural resources in Indiana. He tried hard to be a worthy director.

After leaving the government (at the end of Bush's term in 1993), Ridenour wrote a memoir about his experience. So had two of his predecessors, Conrad L. Wirth and George B. Hartzog Jr., but their works were meant mostly as self-congratulatory vindication of their time in charge. Ridenour included photographs of himself with the president and first lady, and in a majestic natural setting, but he intended to go further and deeper—as readily shown in the title, *The National Parks Compromised: Pork Barrel Politics and America's Treasures*—and he did.

He wrote about his surprise in discovering that the National Park Service wasn't in the executive branch of government:

> Coming from a strong executive branch state like Indiana, I expected to be in charge of running the bureau. Instead I found Congress and, worse yet, congressional staffs running it.
>
> I was absolutely amazed to find staffers on Capitol Hill trying to decide whom I would hire and when I would hire them—even threatening to cut specific people out of the budget unless I cooperated in keeping their pets on the payroll. They are into micro-management big time.

Once a senatorial staff member called Ridenour's executive assistant and chewed him out for twenty minutes. He said his boss, the senator, was angry with Ridenour, and threatened to cut the budget and stop bills the director was pushing. That evening, at a reception, Ridenour ran into the staffer's boss quite by accident.

> I introduced myself to the senator and told him I'd heard he was very angry with the Park Service and me. I said I hoped I could get an appointment to work out his concerns. He looked confused. Finally he said, "Hell, I don't have any idea what this might be about. Forget it—and let me buy you a drink."[4]

In his book, Ridenour took a few shots at the Republicans: "I'm not sure when Congress first started to run the Park Service, but I know the trend accelerated during the Reagan years."[5]

He also took a shot at Secretary of the Interior James G. Watt for failing to acquire a key tract of land outside Yellowstone:

> The range just north of the park was where the bison appeared to like to go. Perhaps that route was bred into them, or the grass looked greener there. This was the old Malcolm Forbes ranch. James Watt had the chance to buy the land pretty cheap during his time as Interior secretary. Instead, much of it was sold to Clare Prophet and the Church Universal and Triumphant.
>
> Secretary Watt's negative decision cost the Park Service . . . with major headaches that have lasted to this day. One, the bison became trespassers as they left the park and headed north. Two, the church, known by the acronym CUT, moved in next door and became a major pain in the rear—not only for the Park Service but for all its neighbors.[6]

But Ridenour was totally nonpartisan in his criticism of Congress:

> The president's [Reagan's] budget request would be low and the members of Congress would start loading up the Park Service's appropriation like a Christmas tree. The pork barrel, or, rather, the park barrel, was so full, it was slopping over. . . .
>
> One of my biggest surprises was to discover what poor condition many of our national parks are in. Roads, trails, employee housing, buildings, sewer systems and other infrastructure are in bad shape. . . . The infrastructure of the national park system is sliding into mediocrity. Congress is quick to give the Park Service new construction projects but loathe to give it the people and the money to take care of its facilities once they are built.[7]

Such is the chronic, painful story of "the national park system sliding into mediocrity," and always promising to do better. Never mind the good intentions in documents like the Vail Agenda, the 1980 State of the Parks report, or the 1998 Natural Resource Challenge. They read well, but when the chips are down, preservation and protection rarely are in the winning hand. The parks are more like pawns in the game of power politics.

A third person who wanted me at Vail, who called to invite me and who then personally paid my way to get there, was L. W. "Bill" Lane, the former publisher of *Sunset* magazine, which his family had owned for many years and then sold to Time Inc. for several hundred million dollars. Lane was basically a California Republican (whom Reagan had appointed ambassador to Australia), but was absolutely nonpartisan in his love for the parks. He told me the conference program was too bureaucratic and uninspiring and that he wanted me to inject energy and spice into it.

It wasn't in keeping with Ronald Reagan and Reaganism. The leading Republican of his day was an actor turned politician with an easy, shallow, common touch that obscured a deep and harsh right-wing outlook. Early in his first term as president, Reagan dismissed the idea of conserving energy, asserting it wouldn't work because Americans would be too cold in winter and too hot in summer. In his mind it was logical to place a clutch of antienvironmentalists in charge of the environment—people like Ann Gorsuch Burford at the Environmental Protection Agency; John B. Crowell Jr. at Agriculture (a logger's lawyer running the national forests and wilderness); and Watt at Interior, plus all the associates and assistants they brought along. Reagan allied himself with the Sagebrush Rebellion, a movement of Western livestock, mining, logging, and real estate interests clamoring for transfer of federal lands to state and private ownership. There were legitimate grievances with bureaucracy, but the Reagan people were less interested in making the federal bureaucracy function properly than in smashing it to smithereens and giving away the pieces.

Watt had worked in Washington for the Chamber of Commerce and the Interior Department, then had established and directed the Mountain States Legal Foundation to fight for free enterprise against the evils of federalism. Under Reagan's presidency, Watt decided to place many jobs in the national parks on contract to private enterprise—in effect moving to "privatize" the parks. From then on the emphasis in park policy and administration shifted. Public parks had long been considered, through administrations of both Democrats and Republicans, to be like art galleries, museums, and libraries, meant to enrich society by enlightening and elevating individuals who come to them. That changed. National parks were increasingly required to pay their own way, setting a path to privatization the Bush administration would later pick up and follow.

As Interior secretary, Watt derided environmentalists as "elitists," "extremists," and "fanatics." "They are political activists, a left-wing cult which seeks to bring down the type of government I believe in," he said once.[8] For Watt it was Americanism versus liberalism and socialism. He was all for morality—fundamentalist Christian morality. When he banned the Beach Boys from the 1983 July Fourth festival on the National Mall, he showed himself so near the deep end that even the Reagans were embarrassed. From start to finish, for federal employees at Interior, the Reagan years were a time of repression, intimidation, and frustration. National Park Service employees were continually belittled and badgered. In September 1983, Watt put his foot in his mouth with a sick joke about the makeup of a coal-leasing panel: "We have every kind of mix you can have. I have a black, I have a woman, two Jews, and a cripple. And we have talent." He was forced to resign.

His successor, William C. Clark, was an improvement. He came to Interior as an old Reagan regular, chief of staff when Reagan was governor. Clark was quiet; he opened the door that Watt had closed to conservationists, calmed things, and kept Interior out of the headlines before the 1984 election. Then he quietly packed his bag, left the hassle behind, and returned to his ranch in California.[9]

While in charge at Interior, Clark appointed another Californian, William Penn Mott, as director of the National Park Service. It seemed like an unlikely choice—considering that Mott was seventy-four years of age—but it was an appealing choice because Mott was widely known and respected in the parks field. He had been a successful director of California state parks under Governor Reagan and was endowed with enthusiasm and energy belying his years. Mott approached his appointment in March 1985 with confidence.

"If President Reagan can run the country at age 74," he said, "I can run his National Park System at age 75." He recalled his productive relationship with Reagan in California, where they agreed the parks would be administered without interference. "You handle the parks, I'll handle the politics," Mott frequently quoted Reagan as saying.[10] His self-assurance and optimism stimulated good feeling, the sense of a shift away from the harsh, myopic mind-set of the Watt crowd. "When in doubt," Mott vowed, as others had before him, "we must err on the side of preservation. Should we subsequently find ourselves wrong, we can always provide for more use."

Mott also reopened doors that Watt had closed. Speaking in June 1985 to a conservation assembly at Yellowstone, he sought support for his new twelve-point plan. The plan was meant for the Park Service, but he wanted the agency to reach out, with such subheadings as: "Effectively share our understanding of critical resource issues with our public(s)," "Increase public understanding of the role and function of the National Park Service," and "Expand the role and involvement of citizen groups at all levels in the National Park Service."

It read well on paper, but little changed. "We're coming up with new ways of doing things. We're action-oriented now," Mott would say, always enthusiastic. "Everybody is excited about the way things are going on here." He felt that his directives were read, believed, and followed to the letter. In an interview in 1986, he said:

> We're telling our interpreters they have a responsibility to talk about environmental issues. We're asking them to understand acid rain, genepools, biotic diversity. The average public reads about these things, but doesn't really know what they're about. Up to this point our interpreters have been told, "Don't talk about anything except national parks." But we're part of a total, interrelated, interdependent world and we've got to talk about all those things.[11]

He said much the same at the Yosemite Centennial Symposium in Concord, California, in October 1990, where I sat next to him on the panel. He still believed interpreters were instructed to brief the public on real issues, and that they were doing so.

But most personnel hunkered down. Despite Mott's best intentions, there was very little citizen involvement, little taking the preservation message to the people and asking their support on critical resource issues. In a message to the ranks (published in the National Park Service's *Courier* in November 1986), Mott exhorted the cause of activism. He identified two types of managers: "passive" and "assertive." Of the former, he wrote, "Such managers travel the road of least resistance. They unflaggingly follow the rules, rarely question decisions handed down, are careful to stay within the mainstream of opinion, and deliver assignments and handle responsibilities in a satisfactory manner. They do little to provide the kind of support, leadership, and creative initiative necessary for an organization to continually grow and develop." His plea brought scant assertiveness in the cause of preservation. Park administrators are trained in technical fields, in which they feel safe. There is no reward for risk taking.

Worst of all, the seeming improvement in attitude around the Interior Department was all illusion. Mott was undermined from the very beginning by his superiors. He had been appointed director of the national parks on Clark's initiative and recommendation.[12] But Donald P. Hodel chose William Horn as assistant secretary in charge of national parks. And Horn, one of the old Watt-Hodel team, went after the Park Service with a vengeance—countermanding Mott's orders, interfering in personnel decisions, changing Mott's recommendations for bonus awards to members of his staff, even changing Mott's efficiency rating of a regional director named Howard Chapman.[13]

Horn was in his mid-thirties—half Mott's age—but that didn't stop him. He overruled Mott's order to restrict aircraft over the Grand Canyon and off-road vehicles and hunting in other park units, true to Watt's pledge to favor public use over resource protection.

Horn had bigger fish to fry. He was the driving force in the Interior Department's recommendation to open the Arctic National Wildlife Refuge in Alaska to oil development. He also negotiated deals in secret with several Native corporations—entrepreneurial syndicates created by Congress with minimal relevance to the traditional Natives—by which the corporations agreed to trade a million acres of remote property for rights to 166,000 acres within the Arctic refuge containing potentially enormous oil deposits. As critics, including the state of Alaska, complained, the federal government would be yielding oil deposits so valuable it would have been feasible and advantageous to purchase the million acres with oil revenues. Luckily, congressional approval was necessary and the secret came out. Horn, meanwhile, in June 1988, a few months before Reagan left office, resigned from Interior to join a law firm representing a number of Native American corporations, including one due to receive oil rights in the Arctic National Wildlife Refuge.[14]

Before departing, Horn ordered a major reorganization of personnel, without Mott's knowledge or approval. One of the changes would have forced the early retirement of Howard Chapman, Western Region director, by downgrading his annual job evaluation from excellent to unsatisfactory. Chapman, normally a cautious administrator, felt it necessary to strike back. He asserted publicly that the move was made against him for resisting the A-76 order for private contracting of park work and for taking environmental positions, and that political appointees were "dismembering the professional capability of the National Park Service." When Chapman retired later, on his own timetable, he left as a hero.[15]

Mott was just getting his second wind. He outlasted the pesky Horn and remained unfazed and optimistic. He was a rare bright light shining through the darkness of the Reagan days. The national parks were hurt, all of them, but, like Russell E. Dickenson before him, Mott kept the parks and the Park Service from being thoroughly debased. Once the Bush administration took hold, however, Mott was dumped. He didn't want to go. He wept in the farewell meeting with close associates after hearing them pay tribute to his sense of humanity and purpose; some of them wept, too.

I knew people like Mott, Dickenson, Chapman, and others in the ranks very well and I respected them. Perhaps that is why I was chosen to speak at Vail—to speak for them. "In cases where parks people do want to protect the sanctuaries in their charge," I declared, "they are thwarted from above." Then I cited specific cases then current:

> Everyone here is aware of how two senior officials were dumped unceremoniously from key positions in charge of national forests and national parks in the Rocky Mountains. They weren't exactly fired, but summarily transferred out of the region; however, as they told a recent congressional hearing, they fell victim to political pressure from Western members of Congress and President Bush's own staff at the White House.
>
> I believe these two career officials and cheer them for going public, as I'm sure that many good civil service professionals do too. Anyone in the Forest Service can easily understand the words of John Mumma, the deposed regional forester, in citing "undue interference and pressure by political figures" for ever larger timber harvests without regard for ecological consequences. Moreover, Secretary of Agriculture [Edward Rell] Madigan recently declared the Forest Service could do "a much better job cutting timber without interference from the courts." He called for an end to the appeals process and judicial review. I find that very scary—a signal that sound professional administration of our public lands, complete with citizen involvement and input, is in serious danger.

> As for the National Park Service, Lorraine Mintzmyer as a regional director testified in Congress that her efforts to prepare a 60-page "vision plan" for the future of the Greater Yellowstone Ecosystem were scuttled through pressure from John Sununu, the president's chief of staff, and subsequently weakened because of "strictly political concerns."

Lorraine Mintzmyer was in the audience. I asked her to stand and be treated to the cheers of her friends and fellows. Mintzmyer had started her Park Service career as a secretary; took various courses; and advanced through the ranks to become a park superintendent, a deputy regional director, and then the first woman regional director in the National Park Service, honored with various meritorious service awards. She and John Mumma should have been honored for preparing a vision plan for the future of the Greater Yellowstone Ecosystem. We are still awaiting such a plan from the National Park Service and our other public agencies.

35

"Their Labors Were Not in Vain"

Through my writing about the politics of parks and the environment in various periodicals, I met members of Congress and became friends with at least a few of them, both Democrats and Republicans. For example, soon after Gaylord Nelson came to Washington in 1962 as a senator from Wisconsin, I called at his office. He welcomed me, sat me down, and brought out a big scrapbook filled with newspaper clippings showing the popularity of his efforts, as governor, to promote Wisconsin state parks. I was always welcome in his office and remember, in 1969, lunching with him in the Senate dining room when he told me of his plans to initiate an event in 1970 to be called Earth Day.

While in Washington, and in travels as well, I observed members of Congress in action. I found that most knew very little about national parks or their purpose and meaning. On the whole, parks have been treated as pure pork and plums for the picking. In the 1960s Julia Butler Hansen of Washington State became chairman of the House appropriations subcommittee controlling the national parks budget. When she visited Glacier Bay in Alaska she told the park superintendent, Robert Howe, that he ought to build roofs over all the lodge walkways so that visitors wouldn't get wet—which is certainly one way to avoid experiencing a rain forest. She also wrote into the budget funding for a reconstruction of Fort Vancouver National Historic Site in her own district, where her son, as it happened, became the curator.

Cecil D. Andrus, secretary of the Interior under President Jimmy Carter, said that most members of Congress were "narrow-sighted and selfish." This may explain why they load the federal budget with park projects for their own districts, sometimes shoving them through appropriations and totally circumventing the prescribed procedure of authorizing first. They will gladly show up at dedication ceremonies and ribbon-cuttings with patriotic posturing and positive words about the parks, while generally ignoring the need to maintain roads, trails, buildings, sewer systems, employee housing, and plumbing.

One day I called on Clark Stratton, the deputy director of the National Park Service, at his office in the Interior Department. He was a friend and made me welcome. He said he had just gotten off the phone with Speaker of the House John McCormack, and this is how the conversation went:

"Yes sir, Mr. Speaker, what can I do for you?"

"My parish priest got a ticket on your Mount Vernon Memorial Parkway. You can fix the ticket. That's what you can do for me."

Many, maybe most, members of Congress would say about the same. To be kind, however, they are not, as a general rule, biologists, ecologists, or historians, either by education or experience. Politics they know better, and are apt to be carried away by it. But I learned that a few have cared deeply about our national parks in a very personal way, and have exercised the political process to help protect and preserve natural, historic, and cultural treasures. They are more fun to write about. And the entire nation has benefited from their efforts.

Congress itself deserves appreciation and at least a round of cheers. It hasn't always done the right thing, but in 1916 it did indeed establish the National Park Service in order to conserve scenery, natural objects, historic objects, and wildlife and to perpetuate these resources for future generations. That 1916 legislation was introduced by Representative William Kent, a prominent California Republican who helped save the redwood grove in Marin County now known as Muir Woods National Monument.

National parks bring out the best in politicians. Senator Harry F. Byrd, a conservative Virginia Democrat, supported national parks throughout the country, but most loved Shenandoah National Park in his own state. As governor of Virginia before coming to the Senate, he worked with citizen advocates and with the legislature to acquire the land for the park. Over the years Senator Byrd climbed every peak in the mountains and his favorite, Old Rag, every year. A crowd always went with him. But at a ceremony in the park, Senator Byrd spoke to posterity when he said, "In the tragedies and other strain of our modern world generations to come will receive peace of mind and new hopes in lifting their eyes to the peaks and canyons of the Shenandoah National Park and those who made possible its establishment can justly feel that their labors were not in vain."

Richard Neuberger felt the same. He arrived in Washington from Oregon in 1954 as a newly elected Democratic senator, a journalist turned politician. He was the foremost media interpreter of environmental issues in the Northwest, contributing regularly to the *New York Times* and major national magazines. He brought with him a strong conservation platform and held to it. Neuberger later recalled advice given to him by an older senator: "Say nothing, and say it well." He did not play it safe, but became "Mr. Conservation," associated with every important piece of environmental legislation before the Senate.

Neuberger believed in the democratic process and loved the Senate. He was never stubborn or bull-headed. When colleagues learned of his cancer, Republican and Democratic senators alike called at his office to inquire after his health. Senator Barry Goldwater and his wife said they were praying for him. He died in 1960 at the age of forty-seven. There is no telling how much good he might have accomplished had he lived.

In the House of Representatives, John P. Saylor, a loyal Republican from western Pennsylvania, championed national parks for more than two decades until his death in 1973. I knew him well and admired him as fearless and tireless, with courage to take strong, forthright, and completely independent positions. He was a tenacious scrapper and never gave up. In 1956 he introduced the Wilderness Act in the House and fought for eight years until it finally became law in 1964. He sponsored and supported the Wild and Scenic Rivers Act, the Land and Water Conservation Fund Act, and the establishment of North Cascades National Park.

Although he received the John Muir Medal from the Sierra Club, "Big John" was also given the Distinguished Service Award from the ultraconservative Americans for Constitutional Action. His motivation likely was derived from old-fashioned patriotism, a belief in America the Beautiful and in Americans as well. He and I were good friends. One day in his office he autographed for me a copy in Latin of what he called "Saylor's Law." The translation began, "If you grab them by the *orchides*"—you can translate that for yourself from Latin—"their hearts and minds will follow."

Another time when he was in Bethesda Naval Hospital, his secretary, Ann Dunbar, telephoned. She said that Saylor wanted me to prepare a statement for him to insert in the Congressional Record condemning recent illegal eagle killings in Wyoming. He was ill, but still alert and championing the wildlife cause. When Saylor was critically ill, Representative Joe Skubitz, a conservative Kansas Republican, pressed in committee to pass legislation Saylor had championed to acquire land on the Maryland side of the Potomac River in order to protect the view from George Washington's home at Mount Vernon. "We must do this for John," said Skubitz, and so it passed.

Phillip Burton was also a patriot, in his way—a two-fisted, fighting San Francisco liberal who stood for every liberal cause in Congress from 1964 until his death in 1983. From 1977 to 1980, while chair of the subcommittee on national parks, he helped to triple the number of miles of national trails, double the number of miles of Wild and Scenic rivers, and pass the Boundary Waters Canoe Area Wilderness Act of 1978, which protects the greatest canoe area on earth.

The day after he died, the House devoted three hours to eulogies from colleagues of both parties and then passed the California Wilderness Act, Burton's legislation to preserve five million acres of wilderness. Burton believed in safeguarding the public's assets as a responsibility of government. He was adamant that the public should never have to pay a fee to enter public parklands and saw any recreation fee test program as a ploy to privatize public lands.

Members of Congress listen to many voices. It is their job to do so. But they benefit when they pay heed to citizens who truly care and want the parks safeguarded rather than exploited. The late Morris K. Udall would attest to that. He was elected to Congress in 1961 and became chairman of the House Interior Committee in 1977. He served until he was taken ill and resigned in 1991.

"Yes, I was on the right side in greatly expanding wilderness and national parks," Udall said before he retired. "But there was one major battle in my own backyard—over two proposed dams on the Colorado River in Grand Canyon National Park—in which, to my friends in the conservation movement, I was one of the bad guys, the tall, lanky fellow in the black hat. This was one of the major environmental battles fought during the 1960s—and, in retrospect, I was on the wrong side. I can only thank God that the growing environmental movement outgunned those of us—including my brother Stewart, then Secretary of the Interior, Senator Barry Goldwater, the rest of the Arizona congressional delegation, and most other western congressmen—who favored the dams."

Fortunately, Udall listened and learned. In 1980 when the Alaska lands bill came to a vote, he said, "I've been through legislation creating a dozen national parks and there's always the same pattern. When you first propose a park, and you visit the area and present the case to the local people, they threaten to hang you. You go back in five years and they think it's the greatest thing that ever happened. You go back in twenty years and they'll probably name a mountain for you."

Then he continued, "In a single vote we doubled the size of the national park system and preserved the 'Crown Jewels' for posterity, in large measure because of the dedication of John Seiberling."

In 1977, as a subcommittee chairman, Seiberling conducted hearings in major cities in the Lower 48, plus remote rural places in Alaska. He poured over maps and listened to testimony from more than 2,300 witnesses—some of whom did not quite think of him as a moderate. Seiberling and Udall held firm against powerful opposition until, as Udall said, Congress effectively doubled the size of the national park system, establishing in Alaska great new parks, plus wildlife refuges, wild and scenic rivers, and major additions to the National Wilderness Preservation System.

Seiberling served in Congress from 1970 to 1987. He was gentle, firm, and forward-looking; he spearheaded legislation to protect architectural, cultural, and historic sites—perhaps because he learned their value growing up in Stan Hywet Hall, now one of the showpiece public properties of Akron, Ohio.

Once, after he retired, he introduced me at a conference. "I'm glad to be on a program with Frome," he said. "He makes me look like a conservative."

Seiberling was followed by Bruce Vento, who began his professional career as a biology teacher in the public school system of St. Paul, Minnesota, unlike many members of Congress who began as lawyers. He was elected to the state legislature in 1971 and, five years later, to Congress. During his twelve terms in Congress he concentrated on safeguarding the public estate. As chair of the natural resource subcommittee on parks, he was responsible for passing more parks legislation than any previous chair.

About the idea of making the parks pay their way, of treating them like a profit-and-loss business, Vento said, "Our national parks were established for people, not profiteers. They are to be preserved and protected. . . . Plans to increase park entrance fees, charge youngsters, and change the collection system in other ways may compromise the access people now enjoy. The main focus of national park policy should be accessibility to the public for a nominal fee, or for free."[1]

Vento died prematurely of a rare blood disease. The following year I was privileged to pay tribute to him at a dinner in Washington, under the auspices of the National Parks Conservation Association. It was an honor for me, particularly with his family from Minnesota present for the event.

The legislators cited above had a direct connection through their committee assignments with national parks. Paul H. Douglas did not. I am sorry I did not know him. He was a World War II Marine veteran and a former professor at the University of Chicago. In May 1958, as the senior senator from Illinois, he stood before a sweep of beach and dune along the southern shore of Lake Michigan, backed by huge sand dunes, between Gary and Michigan City, Indiana, and announced that he would introduce a bill in Congress to make the area a national park. He might have quoted Carl Sandburg, who wrote, "The Dunes are to the Midwest what the Grand Canyon is to Arizona and Yosemite is to California. They constitute a signature of time and eternity." It wasn't easy in an area dominated by steel mills, power plants, and industrial development, but Douglas and supporters insisted these "dunes of incomparable and irreplaceable beauty" must be saved. The act establishing Indiana Dunes National Lakeshore that passed Congress on November 9, 1966, preserves dunes, bogs, oak forest, and marshes, as well as resting, nesting, and wintering areas for many species of birds. It preserves a magnificent fragment of the original America, thanks to Douglas's initiative and energy. And now the Paul H. Douglas Center for Environmental Education provides learning programs for schools throughout the area.

I did, once, meet Charles E. Bennett and remember him as walking with a cane and shaking hands with a smile as politicians do. He was first elected to Congress by the voters of Jacksonville, Florida, in 1949, and was reelected for twenty-one more terms until he retired in 1993. Before his first term he had graduated from the University of Florida law school and served in the state legislature. And before that he had been an army guerrilla fighter in the Philippines, where he contracted polio, leaving his legs paralyzed.

He was respected as a straight arrow by his colleagues, who nicknamed him "Mr. Clean." Though not a trained historian, Bennett was a historical scholar nonetheless, the author of nine books about the history of north Florida. He became especially fascinated with Fort Caroline, the French settlement on the St. Johns River that was wiped out by the Spanish in 1565. This was the subject of a book he wrote published by the University of Florida Press in 1964 and reprinted several times, last in 2006, by the Eastern National Park & Monument Association.

Bennett labored for many years for the protection of the Fort Caroline site and for its reconstruction as the first permanent European settlement in the New World. Fort Caroline National Memorial was established in 1963, but this was not the end. In due course the park was enlarged and the fort became part of the Timucuan Ecological and Historic Preserve, a wonderful area encompassing ancient Native American sites and the Kingsley Plantation, restored to present a picture of nineteenth-century slavery days along the St. Johns River.

Former Congressman Bennett died in Jacksonville in 2003 at the age of ninety-two. Despite his physical infirmity, while in Congress he recorded the longest unbroken string of meeting roll call votes. Perhaps more impressive, he returned his veterans disability pension to the federal treasury and gave his leftover campaign funds to the National Park Service. Over the years he contributed more than $1 million to the Fort Caroline project. Little wonder the *Florida Times-Union*, on September 9, 2003, featured an editorial titled "Charles E. Bennett: A Noble Life."

I might say the same for one and all of the members of Congress whose interest and efforts on behalf of national parks I have described briefly above. Yes, there are those scandals and self-serving deals by congressmen and ex-congressmen cashing in on their connections. But national parks present a better side of America. I think specifically of the Harry S. Truman National Historic Site at Independence, Missouri. It was the only home Truman owned in which he lived and died. Actually, his wife had inherited the house from her mother and, other than their years in the White House, they lived their entire lives there.

When he retired from office in 1952, Truman's income was a U.S. Army pension reported to have been $13,508 a year. Congress, noting that he was paying for his stamps and personally licking them, granted him an "allowance" and, later, a retroactive pension of $25,000 per year. As president, he paid for all of his own travel expenses and food. Later, when Congress was preparing to award him the Medal of Honor on his eighty-seventh birthday, Truman refused to accept it, writing, "I don't consider that I have done anything which should be the reason for any award, Congressional or otherwise." When offered corporate positions at large salaries, he declined, stating, "You don't want me. You want the office of the President, and that doesn't belong to me. It belongs to the American people and it's not for sale." Then he summed it all up: "My choices in life were either to be a piano player in a whore house or a politician. And to tell the truth, there's hardly any difference!"

Back, for a moment, to Gaylord Nelson. Following his defeat after three terms in the Senate, he shopped around for a new career. He found it with the Wilderness Society, advising on policy issues and making speeches here and there. He could have made a lot more peddling contacts and connections, but that was not his goal. In 1995, while I was still teaching at Western Washington University in Bellingham, the Northwest representative of the Wilderness Society telephoned me. He had Nelson in tow in Seattle and wanted to bring him up, if I could get some students together to hear him. The next day we had a healthy turnout and former Senator Nelson delivered an uplifting message. Then we repaired for lunch to a simple off-campus Korean restaurant. It was a far cry from the Senate dining room, but then my friend has his memorials in the Gaylord Nelson Wilderness, the Apostle Islands National Lakeshore, and Governor Nelson State Park at Madison, Wisconsin.

36

Could My Words Possibly Make Any Difference?

On April 7, 1986, I accepted the very first annual Marjory Stoneman Douglas Award, presented by the National Parks Conservation Association (NPCA) at an event in Washington, D.C. Stoneman Douglas stole the show that night and there wasn't a thing I could do about it. As my friend Stewart Brandborg summed it up at breakfast the next morning, "She talked a long time, probably too long, but she's entitled." After all, she was ninety-six at the time.

I sat on the dais between Stoneman Douglas and William Penn Mott, director of the National Park Service. Being next to Mott was an honor in itself. He was a straight arrow who had been director of California state parks—possibly the best state park system in the country—and was then doing his best to protect the national parks from being politicized and poisoned by Reagan administration people running the Interior Department. I had met Stoneman Douglas that day at a lunch hosted by my friend Paul Pritchard, president of the NPCA, exchanging cordial greetings and learning more about her as a person.

Stoneman Douglas spoke first that evening and went on for forty minutes or so. It was inspiring to hear a woman of advanced years express youthful ideas. I remember when she said emphatically, "If the politicians fail to act to protect the Florida panther, I'll get school children all over America to write letters!"

Then Pritchard said kind words in presenting the award to me. I looked out at a friendly crowd of 150 or more public officials and environmentalists, some dear friends and others my sparring partners, all of whom cheered, then cheered again when I said I was giving away the $5,000 award money.

I said that I felt humbled by it all, and hopeful, and was reminded of attending a reception for Ansel Adams when he was about to receive one of the many honors that came to him. I told Adams that it was wonderful his creative genius should be recognized and rewarded. "My mother told me," Adams reflected, "that if I just kept at it, I might amount to something." Gifford Pinchot said it a little differently: "The most powerful thing in human affairs is continuity of purpose."

I recalled that when I was younger, beginning my career as a journalist, I wondered what became of all the words I was cranking out. Did anybody ever read them? Could they possibly make any difference? I discovered in due course that I had chosen the most wonderful way of life. While I was a columnist for *American Forests* I received a letter from a woman engaged in efforts to prevent mining in the mountains bordering her hometown, Colorado Springs. "The most beautiful word in the English language is hope," she wrote, "and you have given meaning to it for us here." A column I wrote became the rallying point for the heroic and ultimately successful crusade to save Overton Park, the beautiful forest in the heart of Memphis. Those words *did* make a difference. During my time as conservation editor of *Field & Stream* I received many, many letters from readers all across America telling me of their special causes and concerns. "Have we so much of earth that we can afford to sacrifice any part of it?" So demanded a woman in western Montana, giving me a challenge I have never been able to put down.

I also thought of another meaningful experience in 1970, when I was invited to Florida to meet with Marjorie Harris Carr, the founder and president of Florida Defenders of the Environment (FDE), and go out with her to view the scene connected with a massive piece of construction called the Cross Florida Barge Canal. The canal was a project of the U.S. Army Corps of Engineers, intended to straighten out the Ocklawaha River, the main tributary of the St. Johns River, and convert it into a commercial waterway linking the Gulf of Mexico with the Atlantic Ocean. At the time of my visit the project was more than twenty-five percent complete at a cost of many millions of dollars.

That did not deter Carr. Her husband, Archie, was a preeminent turtle man—a professor at the University of Florida; author of books about turtles, both scholarly and popular; and a founder and scientific director of the Sea Turtle Conservancy. But Carr stood on her own. She was attractive, matronly, mother as well as wife, well spoken, literate in natural science, and determined.

Years later she would say, "I was concerned about the environment worldwide. How could I affect things in Alaska or the Grand Canyon? But here by God was a piece of Florida—a lovely natural area right in my backyard—that was being threatened for no good reason." FDE brought suit to block completion of the Cross Florida Barge Canal because of the damage it would cause to the geology, hydrology, and ecology of the region. President Nixon learned about it and in January 1971 signed an executive order halting further construction. In 1998 the state dedicated the Marjorie Harris Carr Cross Florida Greenway, lush with nature and absolutely fitting. Harris Carr was already gone to her reward, but her words remain to cheer us, even now: "I am an optimist. I believe that Floridians care about their environment. If they are educated about it, if they are never lied to, they will become stewards of the wild places that are left."

Then there was Marjory Stoneman Douglas. In 1948 Stoneman Douglas sparked a major step forward in public understanding and support of the Everglades with her book *The Everglades: River of Grass*. Before starting the book, Stoneman Douglas knew little about the area. She was writing society news and editorials for the *Miami Herald* (which her father founded) and serving as a columnist on moral issues, condemning prison conditions and urging women's suffrage. She knew her way around Miami but was ignorant of the wilderness at Miami's back door, saying, "When Hervey Allen [New York editor of the "Rivers of America" series] asked me to do this book, I was overwhelmed with the realization that, although I had lived in South Florida for many years and had known some parts of the Everglades, I had no idea of what they were or where I could begin to write about them."[1]

Stoneman Douglas's book awakened fellow Floridians and other Americans to the wonder of the Everglades. Her words still ring like a wake-up call:

> There are no other Everglades in the world. They are, they always have been, one of the unique regions of the earth, remote, never wholly known. Nothing anywhere else is like

> them: their vast glittering openness, wider than the enormous visible round of the horizon, the racing free saltness and sweetness of the massive winds, under the dazzling blue height of space. They are unique also in the simplicity, the diversity, the related harmony in the forms of life they enclose. The miracle of the light pours over the green and brown expanse of saw grass and of water, shining and slow-moving below, the grass and water that is the meaning and central fact of the Everglades of Florida. It is a river of grass.[2]

From that time on, her writing and activism reshaped the public image of the Florida Everglades. With poetry in her prose and spiritual power, she showed the Everglades as more than a swamp to be dredged, filled, and converted into real estate. When she came into my life, many years later, she was still going strong, projecting an image of hope and heart, for the Everglades and all endangered fragments of earth.

Stoneman Douglas went on to found Friends of the Everglades and never quit working to protect the park. In 1991 she celebrated her one-hundredth birthday at a Miami nature center named for her. She was mostly blind and almost deaf but that did not stop her from living in the future rather than the past, or from pledging to keep fighting to save her beloved Florida. "It's the greatest opportunity anyone could have," she told a cheering crowd of friends and fans, then adding, very simply and confidently, "The future is ours." And she kept going on that belief until she died in 1998 at age 108.

In more recent times, the monumental process of rescuing and restoring the Florida Everglades, which Stoneman Douglas championed, has been called the largest, costliest conservation project ever undertaken in this country, if not in the entire world. That may be. Under terms of the Comprehensive Everglades Restoration Plan (CERP) passed by Congress in 2000, it will cost the federal government an estimated $10.7 billion, and the state of Florida $11.8 billion. It will take thirty-five years or more and involves sixty-four major projects, including the purchase of 187,000 acres from the U.S. Sugar Corporation.

The U.S. Sugar property sits between Lake Okeechobee, in central Florida, and Everglades National Park, in the southeast of the state. The property will be used to build and hold massive reservoirs to filter polluted water before it flows to the Everglades and Florida Bay. The land purchase will also enable a halt to the flow of polluted agricultural water now channeled to estuaries on both the Atlantic and Gulf coasts.

It hasn't been smooth or easy. One may ask whether, in this setting, in this age, it is really possible to achieve a comprehensive program to protect and preserve a natural area larger than the state of Delaware. I believe my friend Nathaniel P. Reed summed it up at a hearing before the U.S. Senate Committee on Environment and Public Works at Naples, Florida, on January 7, 2000, when he said:

> Those of us who grew up in Florida in a different era can tell stories of a Florida that barely exists any more. In my lifetime, they have drained and paved South Florida in the name of progress based on a set of values reflective of the time. We now know of the humble consequences that the Everglades has paid, but we are fortunate that we still have a resource that is savable. . . .
>
> We have a rare opportunity to give back to our children and grandchildren an opportunity to experience what my generation of Floridians was fortunate enough to enjoy—a pristine Everglades. It is enormously important to Florida, and it is equally important to the catalog of American treasures that many of us have worked so hard to protect.[3]

Reed has been a good friend since he served as a federal official in Washington (under Presidents Nixon and Ford). He told me then that when he returned home to Florida he would devote his energy to restoration and protection of the Everglades. And so he has, helping to combine and coordinate efforts of Florida environmental groups to press for restoration and raising funds to make those efforts possible. "Nothing like this—a restoration of this size and expense—has ever been attempted," he wrote me. Yet through it all he remained hopeful and optimistic.

These efforts of Reed and many, many Floridians have borne fruit, as evidenced in a December 2011 news report in the South Florida *Sun Sentinel* headlined, "Glades restoration money thrills conservationists." It read:

> Despite the budget crunch here and in Tallahassee, Everglades restoration continues to draw widespread support—and surprisingly generous doses of funding—from Republicans and Democrats in both capitals.

> Though deadlocked on many other issues, Congress provided $142 million for Everglades projects as part of a big spending bill signed into law last week by President Barack Obama.
>
> Gov. Rick Scott, meanwhile, has proposed spending $40 million of state money on the 'Glades during the next fiscal year that begins July 1, more than double the amount he sought for the current year.[4]

Stoneman Douglas should have the last word here. From *The Everglades: River of Grass*: "The future of South Florida, as for all once-beautiful and despoiled areas of our country, lies in aroused and informed public opinion and citizen action."[5]

That tells it in one sentence. Aroused, informed public opinion and citizen action are what it takes. I mentioned above that when I accepted the Marjory Stoneman Douglas Award with personal pride, I could not help sharing the occasion—and the award money—with others who have taken responsibility for safeguarding a special fragment of earth that holds meaning to them. If I had the opportunity to bring it up to date, I would choose the following citizen groups, starting, in tribute to Stoneman Douglas, in Florida:

The Everglades Coalition, composed of fifty environmental and civic groups, has grown in influence and determination. Since 1993 it has been encouraged and aided by the Everglades Foundation, a cluster of wealthy Floridians who have opened their pockets. In three years alone, the foundation gave more than $4 million to sixteen green groups to spend on Everglades issues. Grants are lifeblood for smaller groups such as Everglades Law Center in Fort Lauderdale, whose four attorneys represent environmental groups in lawsuits. A Coalition member group, the South Florida Wildlands Association, reminds us that the Florida panther needs both designation of "critical habitat" and expansion of the Florida Panther National Wildlife Refuge for its survival. Let these actions be considered urgent and taken forthwith.

Buffalo Field Campaign (BFC) has diligently defended the wild bison of Yellowstone, hoping to raise awareness and stop harassment and killing, continually refuting official claims that buffalo transmit brucellosis to cattle, though there has never been a single documented case. It isn't easy while the cattle industry still wields the heavy hand in Montana in the name of "disease management."

The thundering herds will never be heard again, but at least surviving herds yield the feel and flavor of original America. BFC reminds us of how bison help maintain healthy grasslands and drought-resistant plants, of how birds use buffalo fir for nesting, and of how bison calves become prey for wolves and winter kill for grizzly bears and eagles. These are all lessons that should be learned both within Yellowstone and beyond its borders.

Prairie Dog Coalition, now a division of the Humane Society, tends to the well-being of prairie dogs, one of the most loved animal species in national parks, and yet one of the most reviled. On one hand, park visitors are fascinated to watch the dogs—really a rodent, akin to the ground squirrel—atop their burrows, barking and chirping noisily and rhythmically, or emerging from burrows to socialize and feed, bask in the sun, or groom and kiss each other. On the other hand, prairie dogs have been poisoned and shot, and their colonies have been bulldozed to oblivion to make way for "progress." They have been considered pests, or "varmints," by ranchers, farmers, land developers, and allies in government agencies, sometimes including the National Park Service. They have been subject to killing contests with shooters firing mercilessly with high-powered rifles.

The prairie dog (*Cynomys*) is thoroughly American, native to the grasslands of North America, including Mexico and Canada, and nowhere else in the world. Once there were millions, in five different species, living in huge colonies, or prairie dog towns. Now, following decades of persecution, loss of habitat, death from plague, and capture for the pet trade, population has plummeted by 95 percent. Still, luckily, they are found in at least twenty-one National Park Service areas, principally at Badlands and Wind Cave National Parks in South Dakota and Theodore Roosevelt National Park in North Dakota. But it isn't easy for them, even in these refuges.

Biologists and activists associated with the Prairie Dog Coalition remind us of the rich biological diversity associated with the prairie dog, an important prey species: as primary diet of the rare and endangered black-footed ferret, as well as prairie species like the swift fox, golden eagle, badger, and ferruginous hawk. The Coalition hopes to set up demonstration colonies to show nonlethal control methods.[6]

Civil War Trust has warned that Civil War battlefields must not be taken for granted, that many are seriously impacted and threatened by modern intrusion and encroachment. We forget the battlefields were established as national memorials in the post-Civil War era, when the lands around them were essentially farm country and were expected to remain so. The character of the landscape has changed, while basic legislation on land acquisition has not been updated.

Half of Manassas, Virginia, the scene of decisive actions in 1861 and 1863, has been swallowed by residential construction almost in the backyard of the nation's capital. Other Civil War battlefields—Antietam, Fredericksburg, Vicksburg, and Kennesaw Mountain—have also suffered. At Gettysburg, the classic battlefield shrine, much of the nearly 3,400 acres within the authorized boundary is privately owned and subject to commercial development. The field of Pickett's charge is sprinkled with motels and subdivisions. An automobile graveyard flourishes where thousands of Union and Confederate soldiers fought for seven desperate hours, often hand-to-hand, on July 3, 1863. Over the years park personnel have endeavored to remove modern intrusions from the battlefield. The Gettysburg Foundation has raised funds to help underwrite a long-term plan to restore major battle areas to their appearance in 1863.

In 2010 there was a proposal to put a resort casino a half-mile south of the Civil War battlefield on Emmitsburg Road, but the Civil War Trust and other national preservation groups moved to stop the project. The Trust also promotes educational programs and heritage tourism initiatives to inform the public of the war's history and the fundamental conflicts that sparked it.

American Alps Legacy Project, initiated by the North Cascades Conservation Council, is aimed at strong, continuing emphasis on scenic, scientific, recreational, educational, and wilderness values of North Cascades National Park and park expansion in key critical areas.

President Lyndon B. Johnson signed the bill establishing North Cascades National Park on September 3, 1968, but this hardly ended the controversy over the area. Much of it is detailed in *Wilderness Alps: Conservation and Conflict in Washington's North Cascades*, by Harvey Manning.[7]

Friends of Yosemite Valley: When Greg Adair, sparkplug of the Friends, heard the court decision, he exclaimed, "This is a chance for Yosemite to be really remarkable and wonderful!" John Muir surely would have said much the same.[8] The three-judge panel declared the

Park Service's revised plan for the Merced River corridor in Yosemite Valley to be illegal and in violation of the Wild and Scenic Rivers Act.

In handing down the decision on March 28, 2008, Judge Kim Wardlaw for the Ninth Circuit Court of Appeals declared:

> To illustrate the level of degradation already experienced in the Merced and maintained under the regime of interim limits proposed by NPS, we need look no further than the dozens of facilities and services operating within the river corridor, including but not limited to the many swimming pools, tennis courts, mountain sports shops, restaurants, cafeterias, bars, snack stands and other food and beverage services, gift shops, general merchandise stores, an ice-skating rink, an amphitheater, a specialty gift shop, a camp store, an art activity center, rental facilities for bicycles and rafts, skis and other equipment, a golf course, and a dining hall accommodating 70 people. Although recreation is an ORV [outstanding remarkable value] that must be protected and enhanced, see 16 USC 1271, to be included as an ORV, according to NPS itself, a value must be (1) river-related or river-dependent, (2) rare, unique, or exemplary in a regional or national context. The multitude of facilities and services provided at the Merced certainly do not meet the mandatory criteria for inclusion as an ORV. NPS does not explain how maintaining such a status quo in the interim would protect or enhance the river's unique values as required under the WSRA [Wild and Scenic Rivers Act].

The appeals court had addressed the same issue twice before. The appellants, that is, the Park Service, had spent years and considerable public funds to bend the law their way, and had failed again. They had ridiculed the citizen park defenders who ultimately won their day in court.

Park officials grumbled at the decision. Eugene Rose, however, subsequently wrote a little note to enlighten them. It was published as a letter to the editor of the *Fresno Bee*:

> You fail to understand that national parks are not commodities to be bought, sold, or interpreted solely in economic terms. Places such as Yosemite and Yellowstone were not established as moneymakers. They represent the best of the American earth—and under law are to be held inalienable for all times. . . .
>
> Furthermore, it was a federal judge who ruled that the park service had to follow the law concerning the carrying capacity.

> Blame the judiciary or the lawmakers. It was the early conservationists who envisioned the concept of wildland preservation.[9]

Rose has the background to know. Before his retirement, he worked for years as a reporter covering national parks and forests in the California Sierra Nevada and was known as the "conscience of Yosemite." In 1987 he revealed that the Yosemite Park and Curry Company, the park concessionaire, grossed $87 million for its exclusive contract but paid a concession fee of $585,000.

Restore Hetch Hetchy initiated the Yosemite Restoration Campaign in 2012, hoping to ultimately gain public support for draining Hetch Hetchy Valley in Yosemite National Park and developing San Francisco's water supply elsewhere. Before the valley was flooded and the dam built across it, John Muir led a heroic battle to protect Hetch Hetchy—and with it the integrity of all national parks. "It is a wonderfully exact counterpart of the great Yosemite," he wrote, "not only in its crystal river and sublime rocks and waterfalls, but in the gardens, groves, and meadows of its flowery park-like floor. The floor of Yosemite is about 4,000 feet above the sea, the Hetch Hetchy floor about 3,700; the walls of both are of gray granite, rise abruptly out of the flowery grass and groves are sculptured in the same style, and in both every rock is a glacial monument."[10]

Hetch Hetchy Valley is currently under water—water that travels by gravity 160 miles to come out of taps in San Francisco. Restore Hetch Hetchy notes the issue is not simply about water, but about the improved use of local water supplies and the reduction of harm to Yosemite and the wild and scenic Tuolumne River. That makes sense, enabling San Francisco to live up to its reputation for environmental responsibility. The resulting plan would be placed before voters for approval in 2016.

It can be done, as evident in Olympic National Park, Washington, where the Elwha River restoration project began in September 2011. Built in the early 1900s prior to the park's establishment, the 108-foot-high Elwha Dam and 210-foot-high Glines Canyon Dam have blocked the Elwha River's once-legendary salmon runs for nearly a century. Removing the dams will free the Elwha River and allow all five kinds of Pacific salmon, plus steelhead, sea-run cutthroat, and bull trout, to return to more than seventy miles of high-quality habitat protected within Olympic National Park. The return of salmon will restore an important food source for bears, eagles, and other animals, creating a living laboratory where people can watch and learn what happens when salmon return to a still wild and protected ecosystem.

37

When a Whistleblower "Goes Public"

My friend Tony Bevinetto telephoned from Wyoming to ask my advice. The superintendent of Grand Teton National Park, Gary Everhardt, offered him a job as public information officer. I counseled against it, warning that the superintendent might be transferred at any time, leaving Tony holding the bag. Tony did not take my advice and it didn't work that way at all. Everhardt indeed *was* transferred—to Washington, D.C., as director of the National Park Service—and he took Tony with him.

Tony had worked earlier, when we first met, in public relations for the Wyoming Travel Council. I had been out a few years before and we spent a week or so together, touring Wyoming's high spots (mostly on federal lands) and talking with politicians, promoters, and ordinary people. Everybody knew and respected Tony and trusted him as a straight arrow. Little wonder Gary hired him.

Presently, in Washington, I found Everhardt wanted to communicate with me. We met once or twice, then arranged to meet for lunch more or less regularly at one of the better restaurants in Alexandria. He would drive over (or be driven) from the Interior Department and I would come north from my home near Mount Vernon.

Essentially, he wanted me as a sounding board about park issues and personnel. Thus one day he came to talk about filling an upcoming vacancy as deputy director, the number-two position in the agency. "Russ [Dickenson] doesn't want to stay. He wants to go back to the field, and I should let him go," Everhardt said, as though he wanted Dickenson to go. Then he told me what he really had in mind. He wanted to bring in Bill Briggle from Glacier National Park as deputy director, even though (as he said) the regional directors at their recent meeting had urged him not to.

I would have advised likewise. I have mentioned Briggle's name earlier in this book and will again. In an earlier work (*Regreening the National Parks*, University of Arizona Press) I covered the environmental controversy over the boardwalk at Logan Pass, which I will summarize briefly here:

Riley McClelland learned about the proposed boardwalk at Logan Pass for the first time at a meeting of the Glacier resource management committee. He did not initially object to the boardwalk; he was, however, inclined to question the method of finding it the best option, considering that his job was to evaluate the potential impact of proposed changes on park resources. He assumed that the purpose of the meeting was to get input from all those attending. The initial discussion, in fact, centered on whether an environmental impact statement (EIS) should be prepared.

At 6,664 feet on the Continental Divide, Logan Pass has long been a special place in Glacier, the key point of interest on Going-to-the-Sun Road crossing the park between the east and west. For years it was a primitive stop, with a primitive toilet facility. Then, the visitor center was built in the Mission 66 period, followed by a parking lot, water line, and sewage line. More visitors came and toilets malfunctioned. People wandered off marked trails, feathering across the meadows, picking off lichen and chipping rock. Park personnel recommended improving trails with crushed native rock material, which would settle like a driveway.[1] This led to the idea of the boardwalk, which, it was believed, would cause less damage than a trail.

Then Briggle issued a memorandum with his decision that an EIS was not necessary. That has been a very common ploy in ducking diligence throughout the park system. Dr. James R. Habeck, botany professor at the University of Montana in Missoula, called the boardwalk an example of "undue artificiality." "Many alpine ecologists (including myself)," he declared, "believe that building the boardwalk is a terrible mistake, both esthetically and ecologically."[2]

McClelland didn't see any evidence at the time that a boardwalk wouldn't work, but he favored a broad, open evaluation of options. He persisted in questioning the wisdom of these and other developments.

Briggle, however, resisted opposition to his decisions. He was critical of research into the effects of the sewage dump on the bald eagles inside and outside the park. It became known that McClelland's assigned role to bring ecological concerns to the superintendent was meaningless. The controversy led to McClelland's dismissal. He undertook a course of legal action for eight years. It ended in McClelland's favor in a decision by the U.S. Court of Appeals.[3]

In the years that followed, the boardwalk extended two miles to the Hidden Lake Overlook, with faults and failures every foot of the way. In 1985 the park conceded its mistakes via an environmental assessment. The unhappy history at Glacier illustrates the value of openness.

Riley McClelland was a whistle-blower, a federal employee who takes his grievance to the public. That was the cause of his trouble. Montana readers of *Field & Stream* wrote to me more than once about him and I investigated. And then I came to his defense.

That Glacier episode occurred in the 1970s, but similar issues happen every day, even now. Well-meant criticism, from both within and without, is stifled rather than encouraged or heeded, contributing to the National Park Service's difficulties. The bureau hierarchy dreads the critical input of its own people.

Years later, in 1985, I met and interviewed Gary Everhardt at Waynesboro, Virginia. He was no longer Park Service director but superintendent of the Blue Ridge Parkway. He volunteered his views on whistle-blowing:

> I just don't believe in it. My approach is, "If you've got a problem, come talk to me about it. We'll resolve it." Whistleblowing generates a way of saying, "Well, I'm going to squeal on you but I don't have to be confronted with it." I think the accuser ought to stand up and confront the person he's talking about. These people don't seem to be responsible for their actions.[4]

That is one view, reflecting the power of an entrenched bureaucracy. On the other hand, the concerned employee has the legal right to "go public" when he or she feels that internal channels are inadequate. The Civil Service Reform Act of 1978 stipulates that employees are free to make public statements without reprisal, or fear of reprisal, regarding information concerning acts or failures to act by their employer, which they believe harmful to the public interest. The Code of Ethics for Government Service opens with a declaration that "any person in government service should put loyalty to the highest moral principles and to country above loyalty to persons, party or government department." That is the issue here. The National Park Service likes to call itself "a family," and so its personnel are all members of the family and expected to be loyal to it. My experience has shown that whistle-blowers almost always try to resolve issues within the agency before considering going public. Moreover, whistle-blowers often risk their careers, and marriages, by publicly challenging an agency. Blowing the whistle is not undertaken lightly and seldom are there rewards for their actions.

I ask if the internal dissenter must be sidetracked as a "renegade." And if the external critic must be treated as the "enemy." In autumn 1972, Superintendent Briggle might have benefited by heeding the warnings of Habeck, the botany professor, instead of endeavoring to undermine him with a complaint to his superiors at the University of Montana.

Ten years later, when I was driving the Going-to-the-Sun Road from West Glacier to Logan Pass, the day was clear and bright, emboldening the high cliffs. The fifty-mile road between St. Mary on the east side and West Glacier opens the heart of the northern Rockies wilderness to those who might never know it except from pictures. But if the value is inspirational, then it must be greater still without the road.

At NPS headquarters, Director Everhardt brought Deputy Director Briggle and me together. I believe we both put the past behind us and sought to move ahead with mutual respect. But there came a time when I was visiting in his office and was privy to him unmercifully chewing out a member of his staff. When that poor fellow withdrew, I couldn't help saying, "Bill, I feel embarrassed to be in your office and listen to you treat an associate so harshly." Briggle said nothing in reply, but I remember that as a tragic scene.

Tony Bevinetto, however, brought good feeling wherever he went. I would visit him and his family at their home in northern Virginia. Then he and I would go off to ski over hills and slopes of nearby Manassas Battlefield. Tony was neither tall nor athletic, but managed well. In due course he left the Park Service to serve on the staff of the U.S. Senate Committee on Energy and Natural Resources from 1979 until his death (of cancer) in 1988. The Park Service soon after established the

Bevinetto Congressional Fellowship to improve mutual understanding and cooperation between the Park Service and Congress. The honor was well deserved.

But this chapter is about whistleblowing. I deal elsewhere in the book with the issues at Hubbell Trading Post in New Mexico and with Chief Chambers of the Park Police in Washington, D.C., and in this chapter with the shabby treatment accorded to Riley McClelland at Glacier. Now I turn to the very painful showdown at the Chesapeake & Ohio Canal National Historical Park.

In my recount, it opens in 1999 when Robert M. Danno received the Interior Department Award for Distinguished Service. As chief ranger of Saguaro National Park in Arizona, he rescued three Arizona Game and Fish Department employees who were stranded on a steep cliff in a rugged area of the Coronado National Forest. Danno rappelled down on an unfamiliar, badly eroded, soft rock face with only headlamp visibility. He found the first victim and lowered him to where the other victims were stranded. The desperate victims feared they could not make it through the night on the cliff, citing fatigue and cold. Danno reassured them and the three were lowered six hundred feet in two stages to the cliff bottom. The entire rescue took ten hours and was complicated by unfamiliar terrain, extremely hazardous cliffs, and poor visibility.

In due course Danno was transferred to serve as chief ranger of the Chesapeake & Ohio Canal National Historical Park outside Washington, D.C. In this capacity, in May 2005, he informed the Office of the Inspector General (OIG) at the Interior Department that something corrupt was afoot along the C&O Canal, specifically that Superintendent Kevin Brandt had allowed Daniel Snyder, the wealthy owner of the Washington Redskins, to remove 130 trees on National Park Service lands held in a scenic easement after Mr. Snyder had offered to make thousands of dollars available for NPS projects.

The OIG investigated and discovered that, in 2002, Daniel Snyder offered $25,000 to the NPS. But the park superintendent at that time wrote that he could not accept this "generous offer . . . as mitigation for scenic easement variance requests." However, Special Assistant to National Park Service Director P. Daniel Smith told the OIG that his boss, then-NPS Director Fran Mainella, wanted the Snyder tree-cutting issue resolved. Smith and the new C&O superintendent, Kevin

Brandt, met at Mr. Snyder's residence to discuss a resolution to the tree-cutting issue.

Superintendent Brandt later told investigators that he was a new superintendent in 2004 and wanted to be considered a "team player." Snyder saw that the remaining exotic and native trees on the easement were cut down in November 2004. Six months later, Chief Ranger Rob Danno notified the OIG.

This was not the first time Assistant Director Smith allowed trees to be removed from NPS properties to improve the views from homes owned by people with power and influence. The OIG investigation concluded that Smith had violated NPS policy and procedures when he influenced the Snyder tree removal approval.

OIG, in a letter to Lynn Scarlet (Acting Interior Secretary), said:

> Investigation determined that NPS failed to follow any of its established policies and procedures outlined in the NPS Director's Handbook, and even disregarded the recommendations of their own Horticulture Advisory and Review Committee, regarding the process in which a property owner on an NPS scenic easement can cut vegetation above the allowable limit.
>
> Specifically, the NPS National Capital Region officials and C&O NHP employees failed to initiate the requisite environmental assessment, as required by NPS guidance, when instituting changes to an easement agreement. In addition, NPS did not complete the required paperwork detailing the reasons for granting Mr. Snyder exclusions to a Special Use Permit, which allowed him to cut vegetation beyond the allowable limit.

P. Daniel Smith was reprimanded in a letter. He later became superintendent of Colonial National Historic Park.

Kevin Brandt remained as superintendent of C&O Canal. Dan Snyder was given a much better view of the Potomac River from his residence, but the hillside where he removed the trees began to erode in 2006. Maryland officials fined Daniel Snyder $37,000 for violating county forest conservation law and ordered him to plant more trees.

Rob Danno, for his effort to follow NPS rules and regulations, was reassigned to the George Washington Memorial Parkway—to issue picnic permits. Then, in 2007, a search warrant was executed on Danno's government-owned residence. Officials found the following "contraband": an NPS drill, an NPS badge collection, a slide projector and accessories, various NPS signs, and an emergency services trauma kit. The U.S. Attorney's Office saw "prosecutorial merit" to the case and indicted Danno. The NPS placed the chief ranger on unpaid status. After a three-day trial, the jury found the chief ranger innocent.

Ranger Danno's attorney, Peter H. Noone, said, "Soon after Mr. Danno made his disclosures, he experienced a variety of administrative actions, including temporary reassignment, investigation, frivolous administrative charges, Board of Inquiry, suspension, isolation, permanent reassignment, and criminal charges."

Mr. Noone added, "Mr. Danno's protected disclosures to the DOI Inspector General were confirmed to be truthful, accurate, and instrumental in assisting with the government's investigation, as well as in assuring the protection of NPS resources. . . . Mr. Danno's protected disclosures [showed] dedication to its agency resources . . . and we believe that he should have been embraced for his courage to come forward."

I noted and became interested in this case. On March 27, 2009, I received the following email from Don Castleberry, who had retired from the Park Service as a regional director:

> As Superintendent of George Washington Memorial Parkway, 1977–1980 (next door to C&O), it was a constant threat, to have wealthy, well-connected neighbors attempting to do selective cutting, to improve the view from their properties, into the river valley. Sometimes, they would sneak out, in the dark of night, and do it over time so it was not immediately obvious. In other cases, they would just use direct political pressure, which came down from "above" and was usually manifested by some political appointee (a deputy assistant secretary, often), taking charge of the matter. They would use scare tactics, threats, subtle persuasion, just short of an action which could be reported by a whistle blower. A superintendent or chief ranger would often find himself in the presence of highly influential "celebrities" of Washington society in an attempt to "persuade" him that he's being unreasonable and might face, unstated, blemishes to his reputation or career if he is not a "team player." In spite of that, it's usually pretty easy to just "stand pat" and refuse, and they would back off. Not so easy if you are new, unsure of the support from your above, with a "hostile" Administration, etc. We did, successfully, prosecute some of those cases on GWMP during my tenure there.
>
> I know almost all the players in this case. Don't know what role Fran [Mainella] may or may not have played, so can't

> comment on her possible action or influence. It appears to me, though, that the rest of this report rings true. I have no doubt that Danny "overstepped his discretion," and was, in fact, guilty of inappropriate political interference. Kevin might have just stood his ground and refused, but I wasn't there and can easily understand him being intimidated, as a new superintendent. I'm not inclined to criticize his actions. Danno did the right thing, paid a price but at least was cleared of the politically motivated charges and the matter came to light. It's not a pretty picture, but I think folks familiar with the Washington, D.C., scene will recognize the pattern.

Danno spells it all out in a book titled *Worth Fighting For: A Park Ranger's Unexpected Battle*, published by Honor Code Publishing in 2013.

But there is more to it. The *Washington Post*, on October 3, 2013, carried a story by Miranda Spivack headlined "Whistleblower in Snyder tree case moves on to a new job, wins settlement with park service." It opens as follows:

> The federal government has settled whistleblower retaliation complaints from a former C&O Canal chief ranger who said he suffered years of reprisals after revealing that the National Park Service had allowed Washington Redskins owner Daniel M. Snyder to cut down 130 trees.

Danno said, "I hope that my experience helps the National Park Service get back on course." He thanked the Public Employees for Environmental Responsibility (PEER) for its efforts in his behalf. As part of the settlement, Danno was ordered reinstated with assignment as a division chief for wilderness planning at the Park Service's wilderness training center in Missoula, Montana.

38

Citizens Challenge the Parkway Center

On October 5, 1998, a citizen organization, the Public Parks Task Force, convened a forum in Asheville, North Carolina, to vent deep distress and lack of input or consideration in the proposed construction of a Blue Ridge Parkway headquarters and visitor center. Superintendent Gary Everhardt, a former director of the National Park Service, was invited to the meeting, but sent a staff member to take the heat of concern and opposition.

I was invited and I came. But first, a bonus of the trip was an advance side excursion to Flat Rock, fifteen or twenty miles from Asheville, to visit Connemara, the home and farm where Carl Sandburg spent the last twenty-two years of his life. It was too good and too close to miss.

Thanks to designation as a national historic site, the property has remained largely as Sandburg and his wife left it, reflecting his immersion in poetry and history and her vocation as a breeder of prize goats. At the gift shop, I bought a copy of Sandburg's *The People, Yes* and read into it that night. It reminded me that Sandburg was really a journalist who collected Americana and chronicled the lives of Americans in his own way.

My thoughts then turned from Sandburg to Walt Whitman. He, too, started as a journalist, becoming, in time, the prophet of liberty through poetry of his own structure. Whitman published the first edition of *Leaves of Grass* at his own expense and never wavered from his chosen course. Once I visited and wrote about his two-story clapboard house in Camden, New Jersey, the only home he ever owned and where he died in 1892. He called it his "little shanty," which seemed about right, since the house was cluttered and run down, in a rundown neighborhood. But that was pure Whitman, the "Poet of Democracy." As he wrote in *Leaves of Grass*, "Who, constructing the house of himself, not for a day, but for all time, sees races, eras, dates, generations, the past, the future, dwelling there, like space, inseparable together."

I asked myself why I was invited to speak at the Public Parks Task Force's forum in Asheville and remembered that I was known in the area through my book on the Smoky Mountains and also through my experience in national parks. Nor was it my first engagement in this region. Following publication of my book *Strangers in High Places: The Story of the Great Smoky Mountains*, I came to receive the Thomas Wolfe Memorial Literary Award from the Western North Carolina Historical Association. In June 1975 I was the keynote speaker at the Fontana Conservation Roundup, a broadly inclusive regional conservation assembly. I spoke on "protecting the public options" with this opening sentence: "There is always hope as long as there is a cadre of people, or even a single individual, with commitment and a willingness to fight."

On that trip to Asheville in the fall of 1998, I flew from Morelia, Mexico (where I was teaching in a foreign studies program), an ancient place grown into a city of one million-plus, laboring under a thick, dark cloud of polluted air. In Asheville, it was much the same; I felt I had hardly left Mexico. But I remembered Asheville a half-century earlier as a charming, healthful mountain resort. Wasn't that what led people to move here in retirement from New England, the Midwest, and Florida? But who was minding the store to protect the qualities they came for?

I suppose in a democracy we the people get no less than we demand and deserve. On the positive side, at the outskirts of Asheville, the North Carolina Arboretum now covered 426 acres of gardens, greenhouse, and educational displays as "a natural cradle of plant cultivation." For the hopeful, there always is hope. But I could not be satisfied with the process of decision-making or the policies governing land use in the Great Smoky Mountains National Park or the Blue Ridge Parkway.

It takes agencies repeated lessons before they learn. In 1974 the Park Service decided to proceed with the access road into Cataloochee Valley, a lovely section of the Great Smoky Mountains park where structures of the old pioneer backwoods settlement still remained. In order to shortcut the requirements of the National Environmental Policy Act (NEPA), officials prepared an "Environmental Assessment" instead, based on their intent to proceed with construction within thirty days. They tried to keep it quiet, never notifying anyone on the Tennessee side of the park, let alone any national publication, national organization, or concerned citizens. Someone in Haywood County appealed to me out of desperation, and I did what I could to spread the alarm. It frightens me that this is how our public officials do business on our public land.

The Environmental Assessment presented valid reasons why the access road should not be built: a potential increase in use from 250 to 8,000 persons per day, a volume which the fragile valley cannot stand; potential for strong temperature inversions; inevitable soil erosion and stream siltation; and inevitable destruction of wildlife habitat. Nonetheless, the agency decided to proceed. When I raised the question with the park superintendent of the potential for a temperature inversion to trap and concentrate auto emissions, as mentioned in the assessment, his response was, "That's just a potential. It doesn't say it's going to happen." If this is the kind of principle guiding our national parks, we are in deep trouble.

A group of citizens in North Carolina and Tennessee decided to hike en masse to protest the proposed road. The park superintendent, Vincent Ellis, a courtly silver-haired gentleman, agreed to meet with the group to explain the park's plans. He clearly was following orders from the regional and Washington offices, designed strictly to accommodate the local congressman, Roy Taylor, who held the key post of chairman of the parks subcommittee in the House. Ellis came to the appointed rendezvous with an entourage and maps. When he was through explaining things, the superintendent said, "Okay, now let's get in the cars to drive around and see where the new road would go." When members of the group insisted they wanted to walk, Ellis was taken aback. "But you won't see nearly as much." But the citizens hiked anyway, and the Park Service people did too, for a change.

When Ellis retired in 1975 he was succeeded as superintendent by Boyd Evison, who had quite different ideas. Evison was someone I knew throughout his entire career and whom I cite as one who made his way without acquiescence to conformity and opportunism. Instead of building new roads, he closed old roads, turning them into what he called "quiet walkways." Instead of garbage cans at scenic overlooks along park roads—facilities that spawn waste and ugliness—he shifted litter collection stations to park exits, where visitors were urged to deposit their trash.

Evison's behavior rubbed some people the wrong way. Park personnel didn't like the idea of traveling in the backcountry by shank's mare where they formerly had driven over "administrative roads," or of using hand tools instead of power machines.

More serious, a handful of politically privileged North Carolinians balked and squawked when Evison closed a cozy fishing retreat maintained with public funds for their private benefit. They were mostly local businessmen and boosters, "civic leaders" interested in the park to the extent that it benefited them, and they had direct lines of communication to their congressman and governor. Then, in 1977, Evison engaged a team of professional hunters to undertake a test project aimed at reducing the number of wild boars uprooting and destroying vegetation in the park. Local shooters protested loudly that *they* had been hunting boars in and around the park, legally and otherwise, for years. The Interior Department came down on the side of the hunters and ordered the project halted. Evison was removed. He survived and later advanced elsewhere, but the message to park superintendents was clear: Do not stand tough on principle. If you want to get along, learn to go along.

I had known Gary Everhardt in Washington in his previous role as director of the National Park Service. He had begun his career as an engineer, working up from there, becoming a protégé of National Park Service Director George B. Hartzog Jr., then assistant superintendent of Yellowstone National Park and superintendent of Grand Teton National Park. Hartzog was already gone when Everhardt was surprisingly chosen to become director in 1977. Several years later, after he had been fired, he told me, "I had never worked in the Washington office. To reach down and get somebody from within the ranks, as in my case, shows a certain amount of naiveté. It's not like being out in Wyoming. It's as different as night and day."

That was the way it went. To politicians in power, national parks and park personnel were (and remain) like checkers on a checkerboard, to be moved, or removed, as part of a bigger game. After Richard Nixon was reelected president, he personally ordered Hartzog's dismissal and the appointment of Ronald H. Walker to succeed him. Walker had neither experience nor desire for the position, but was a good soldier and did his best, until Nixon resigned, spurring Democrats in Congress to get rid of him. Many National Park Service directors were selected out of the blue, lasted two years or so, then left through the revolving door.

When Everhardt became director, one of his first moves was to appoint another Hartzog protégé, William J. Briggle, as his deputy director. This was not a popular move, since Briggle, then superintendent of Glacier National Park, was feared and disliked as a heavy-handed authoritarian. When Everhardt got fired, Briggle went with him.

Everhardt told me that he and Briggle were sent to an office in northern Virginia for about six months. During their exile they stewed in their own juices, with nothing to do except pick up their paychecks. Finally they got their orders: Briggle to be superintendent of Mount Rainier National Park in Washington State, and Everhardt to be superintendent of the Blue Ridge Parkway.

As a native Carolinian, Everhardt didn't mind the assignment, and welcomed the opportunity to return "down home." Once in place, he settled in to stay, overlooking or ignoring an unwritten agency rule that a superintendent remains in place only three or four years to prevent him or her from growing too cozy with local interests. Everhardt stayed as superintendent on the parkway until he retired twenty years later.

During that time the magic mountain vistas were diminished by smoke and haze, the result of toxic emissions airborne from distant factories and from power plants and automobiles. Condominiums and commercial developments of one kind and another emerged on private lands bordering the parkway, while on the parkway the concessions grew bigger but not necessarily better.

Superintendent Everhardt proved more attentive to power and political constituency than to concerns of citizen conservationists. He hosted insiders at parties with liquid refreshment and wild game at the Soco House, a retreat west of Asheville on the parkway, served by park staff personnel. This led to a proposal to construct a substantial headquarters complex and economic development center in ecologically sensitive terrain, with more response to tourism than to ecology. Thanks to the influence of the local congressman, Charles Taylor, an ultra-right-winger, that center was built anyway, and named the Gary

E. Everhardt Headquarters Building. Today, that headquarters building is used for park offices, as it should be, but on weekends and holidays, the entrance gate is locked to the public.

Among those opposing the visitor center was the late Art Allen, who first worked on the parkway as a ranger in 1959, then returned as assistant superintendent in 1981 and retired in 1990. He wrote to me, via email, in 2007:

> The Visitor Center at Hemphill Knob was a real waste of money and damage to the environment. You will probably not be surprised to learn that it is not getting much use, and there are no funds to maintain the facility. It has now been open for more than a year.
>
> I opposed the project from a slightly different angle. I have been a member of the Southern Highland Craft Guild (formerly the Southern Highland Handicraft Guild) for nearly forty years as a stained glass craftsman. I have been president of the Guild for two terms and a board member for three terms. So when the project was under consideration, I wrote the following article in the Asheville paper. The project received more than 90 percent negative comments, but of course, that was to no avail, because when a local politico wants something, it will get done. By the way, the final price tag was over $12 million dollars.

That article, published as a "guest commentary" in the *Asheville Citizen-Times* of October 11, 2004, was headlined, "Taylor's proposed parkway visitors' center a waste of money; a better plan is to expand Folk Art Center." In it, Allen wrote:

> The Southern Highland Craft Guild, which operates the Folk Art Center, began its partnership with the Blue Ridge Parkway prior to the construction of the parkway. In 1937, Guild officials met with the director of the National Park Service to forge a working arrangement that would benefit the public and the craftspeople of the Southern Highlands. That partnership has endured for more than sixty-six years. . . .
>
> If he pursues his unexplained interest in locating a competing visitors' center at Hemphill Knob, he will only achieve irreversible harm to both the parkway and many craft people within his district.

True enough, the parkway and craft guild grew up together, without dollar signs attached. I hope they will find their way back on that track.

39

Horace Albright Typing on His Aged Portable

I called on my friend Horace Albright in 1984 when he was ninety-five years of age, living in a convalescent home in Los Angeles near the home of his daughter Marian Albright Schenck, and her husband, Roswell. I found Albright physically frail but lively, in good spirit. After all, he was a legend to park professionals and citizen admirers like me.

He showed me that he had been typing on his aged portable, essentially hunt-and-peck, writing mostly to widows of old National Park Service colleagues, and to old colleagues, survivors like himself. We talked for at least an hour. For a while he expounded on his conferences with Secretary of the Interior Harold L. Ickes during the New Deal days. It was all very vivid as though it was yesterday. I was delighted to hear it, but interrupted. I pointed to a door in the room, asked if it led to the bathroom and if I could use it. "Why, I'd be *honored* if you would," was his reply. That was vintage Albright.

I was en route at the time to deliver the 1984 Horace M. Albright Lecture in Conservation at the University of California, Berkeley, but I could not miss the opportunity to be with Albright himself one more time. To illustrate what he meant to me, for a period of five or six years, starting in 1977, when I wrote a column called "Environmental Trails" each Sunday in the *Los Angeles Times*, he would clip and send to me a copy of the column with a comment or reflection. I wanted to provide him with stamps, but he brushed me off, saying that would spoil it all.

A few excerpts from his notes to me may be fitting here:

> June 6, 1978: I well remember that when I was Director, I spent two weeks in the Great Smokies, riding horseback everywhere, and I remember Gatlinburg. On returning to Knoxville, I remember publicly declaring that Gatlinburg was the ideal national park gateway town, and compared its beauty, serenity, good taste, etc., with gateway towns and cities in the West—Estes Park, Colorado; West Yellowstone and Gardiner, Montana, etc. A few years later, Gatlinburg had "gone over the dam." I could not say anything good about it. I last saw it in 1961. I feel about it like I do Lake Tahoe—I never want to see it again.

> July 9, 1978: I have always been a train fan. What beautiful trains we had going to national parks in the 1920s to 1950s. The North Coast Limited from Chicago to Yellowstone on Northern Pacific; the Yellowstone Special between Salt Lake City and West Yellowstone; the pleasant Burlington from Chicago to Cody, Wyoming—all Yellowstone trains. In the 1920s these trains would deliver to park entrances over forty thousand happy travelers who toured the park in open buses—except when raining; canvas tops quickly opened and the seated passengers were thus protected. Then there were the Empire Builder and Oriental Limited to Glacier Park; the Grand Canyon Limited on the Santa Fe from Chicago to Grand Canyon; Chicago, Milwaukee & St. Paul to near Mount Rainier, etc.
>
> With officers of the Denver & Rio Grande Western I once covered the entire system, including narrow-gauge high in the southern Rockies; and in 1947 when my U.S. Potash Company was the second biggest shipper on the Santa Fe Railway, I was out with the directors and officers in a special train, inspecting every mile of the system.

> July 23, 1978 (on Eureka Valley, California): Had I known about it, or had anyone told me about it back in 1933 when I was working on boundaries for Death Valley National Monument, I most certainly would have included it—and I could have done it easily. Another thing about your story—you mentioned the "Saline Salt Tram." My father supervised the building of that famous tram. He felt there was no way for him to have a part in what looked like a profitable enterprise, except by undertaking the top supervising job. He did this for a bare living wage, taking stock in the tram company. I was in college at the time, and the sacrifices my Dad had to make left him

very little for me—$200 for my room for my freshman year and none afterwards. Then the tram failed and nobody got anything out of it.

July 23, 1979: The story enclosed revived memories of the Big Thompson Reclamation Project which contemplated moving water from the upper Colorado River in a tunnel UNDER Rocky Mountain National Park at just about the area of the Rockies where the mountains were supreme—their massiveness and power—just where the Front Range branches out from the main range at Long's Peak. We opposed the project and refused to permit surveys to gather data for design of the tunnel and other works. Objections on two grounds—invasion of the park by a commercial enterprise with problems of waste disposal from tunnel building, etc., and diversion of Colorado River waters from west of the Continental Divide to areas east of it, when all water originating on the west side would be needed on that side from Colorado to California. Both Director [Stephen T.] Mather and I succeeded in holding the Big Thompson project in check. When Hoover was president I thought for a while we would be defeated, for the proponents of the project hired a well-known engineer, a classmate of Hoover's at Stanford, to secure permission to make the necessary surveys and design the remaining details of the project.

However, I succeeded in my refusal to make any of these concessions to the engineer, whose name escapes me, and I was never asked any questions from the White House. Along comes the New Deal and Harold L. Ickes. One of the first reclamation projects brought up in the new administration was the Big Thompson. I earnestly recommended against any surveys affecting national park land, and I don't believe Secretary Ickes yielded while I was still there. But before 1933 ended, he authorized the surveys; the project proved feasible and was built with the tunnel under Rocky Mountain Park at its scenic climax. Moral of the tale, don't let any surveys be made in a national park.

June 1, 1984: Since you were in California, we have lost Ansel Adams, as you surely have been advised. He was eighty-two. I knew him from about 1925 and was at his wedding in 1928. He was one of the greatest of photographer-artists, but he was erratic at times in his positions on resource conservation. I was not at all pleased with his tirade against Ronald Reagan. It was uncalled for, and certainly hurt him. His dislike of Reagan was due to the appointment of [Secretary of the Interior James

> G.] Watt. He went a lot too far when he said, after the President spent a half hour listening to him, "I hate Reagan." So far as the Park Service and wilderness were concerned, Watt was fair, although I never thought he was a man for secretary of the interior.

I saw Albright again the following day. His daughter and son-in-law picked him up and brought him to their home. I interviewed him with a tape recorder and felt he was filled with lively recollections. I probably could have listened to him all day and posed still more questions. I hated to have our time together come to an end, but ultimately he returned to the nursing home and I drove on my way north.

I took another side trip to Mineral King, a glacial valley in the southern Sierra, a very recent, major, and contentious addition to Sequoia National Park. Books had shown that miners in the nineteenth century swarmed over Mineral King in their search for silver. For a few short years in the late 1870s it promised to become another Comstock Lode, a mountain of rich silver ore. But harsh times turned the boom to bust; the workings were abandoned and Mineral King became national forestland. In time it grew to be one of the most popular recreation spots in the Sierra. And so in 1965 the Walt Disney Company proposed to transform Mineral King into a massive ski resort as large as Sun Valley, Idaho.

The state of California (with Governor Ronald Reagan), the Forest Service, and local media boosters lined up in support of the ambitious Disney proposal. The Sierra Club led a coalition in opposition, turning to federal courts and ultimately the U.S. Supreme Court. In 1974 the Court ruled against the club, declaring that it "lacked standing," that it was seeking to vindicate its own value preferences through the judicial process. However, this case, *Sierra Club v. Morton*, is most widely known for the dissenting opinion by William O. Douglas.

A word here about Justice Douglas. He was a Northwest-bred outdoorsman who hiked, rode horseback, camped, and fished; a committed conservationist; and prolific author (of *A Wilderness Bill of Rights* and other works). Justice Douglas epitomized bold intellectuality, which led detractors to detest his far-flung activism. Why, they demanded, couldn't he exercise judicial restraint and propriety? He brushed them off, unbending: "A man or woman who becomes a justice should try to stay alive; a lifetime diet of the law turns most judges into dull, dry husks."

An excerpt from his Mineral King dissent:

> The critical question of "standing" would be simplified and also put neatly in focus if we fashioned a federal rule that allowed environmental issues to be litigated before federal agencies or federal courts in the name of the inanimate object about to be despoiled, defaced, or invaded by roads and bulldozers and where injury is the subject of public outrage. Contemporary public concern for protecting nature's ecological equilibrium should lead to the conferral of standing upon environmental objects to sue for their own preservation. This suit would therefore be more properly labeled as Mineral King v. Morton.
>
> Inanimate objects are sometimes parties in litigation. A ship has a legal personality, a fiction found useful for maritime purposes. The corporation soul—a creature of ecclesiastical law—is an acceptable adversary and large fortunes ride on its cases. The ordinary corporation is a "person" for purposes of the adjudicatory processes, whether it represents proprietary, spiritual, aesthetic, or charitable causes.
>
> So it should be as respects valleys, alpine meadows, rivers, lakes, estuaries, beaches, ridges, groves of trees, swampland, or even air that feels the destructive pressures of modern technology and modern life. The river, for example, is the living symbol of all the life it sustains or nourishes—fish, aquatic insects, water ouzels, otter, fisher, deer, elk, bear, and all other animals, including man, who are dependent on it or who enjoy it for its sight, its sound, or its life. The river as plaintiff speaks for the ecological unit of life that is part of it. Those people who have a meaningful relation to that body of water—whether it be a fisherman, a canoeist, a zoologist, or a logger—must be able to speak for the values which the river represents and which are threatened with destruction. . . .
>
> The voice of the inanimate object, therefore, should not be stilled. That does not mean that the judiciary takes over the managerial functions from the federal agency. It merely means that before these priceless bits of Americana (such as a valley, an alpine meadow, a river, or a lake) are forever lost or are so transformed as to be reduced to the eventual rubble of our urban environment, the voice of the existing beneficiaries of these environmental wonders should be heard.

Although the Sierra Club lost the case in court, the club won the war to protect Mineral King over the long run. After years of legal battle, the ski resort was never built. In 1978 Congress voted to annex Mineral King Valley into Sequoia National Park. And that was what led me on my path.

I met my friend Marvin Jensen at park headquarters at Three Rivers and visited awhile. Then we headed by car for the trailhead at Mineral King. Several years earlier when Jensen was stationed at Grand Canyon National Park, I traveled with him by boat down the Colorado River. He later served as superintendent at Glacier Bay National Park in Alaska. He was agile, tireless, and a good friend.

It was only twenty-five miles to the trailhead, but a long drive, winding and climbing along the East Fork of the Kaweah River, on a road built by miners more than a century ago. We stopped for lunch at Silver City, a little settlement at 6,900 feet, a gateway to the high valley. Then, over the next three days, I found what I was looking for—a thoroughly challenging, stimulating, and rewarding national park experience, hiking and backpacking in this area only recently added to the national park. That may sound a little strange, considering that Sequoia, the second oldest national park (second only to Yellowstone) was established in 1890. Both Sequoia and adjacent Kings Canyon were set aside to preserve groves of giant sequoia trees, the largest of living things, remnants of forests that once covered a widespread portion of the northern hemisphere.

The two parks, which are administered as one, do much more. Stretching eighty-five miles from north to south, they protect an unbroken wilderness of granite peaks, gorges, rockbound glacial lakes, flowering alpine meadows, and virgin forests. The parks comprise, for me, the heart of the Sierra Nevada, the country's highest mountain range outside of Alaska, a jagged crest extending from the Cascades southward for some four hundred miles, with elevations ranging from 7,000 to 14,000 feet.

Once we took off from the Mineral King trailhead at 7,500 feet, the steep climb, high elevation, and loaded pack meant little to Jensen. I was in good shape for my age (then sixty-four). With Jensen's encouragement and trust, I surprised myself with my ability to keep up. There was no easy route to the alpine lakes basin, but Jensen chose the toughest unmaintained trail. We reached the first lake, surrounded by a red fir forest, and kept climbing the chaparral slopes flecked with wildflowers, finally stopping to camp on the rocks overlooking the second of three lakes.

The next morning, Jensen proposed climbing the trailhead flank of the high ridge between the Mosquito Lake Basin and the Eagle Lake Range. It was a forbidding path that I would never attempt on my own; with a patient guide and friend, however, I dared to meet the challenge. Besides, the country was irresistible—with granite peaks, gorges, canyons, and meadows, expansive and wild. Once on top, we dropped our packs and advanced up the rocky, knife-edged ridge. Jensen indeed knew where he was going—we intercepted the trail to Eagle Lake at about 11,000 feet. Descending from the ridge on the Eagle Lake side, after picking up our packs, should have been easier, but the slope was steep and rocky, with narrow ledges and loose, sliding material. Yet there was chaparral in the rock, abundant with blooming wildflowers, including *rein orchis*, Indian paintbrush, shooting stars, and leopard lily.

During our pilgrimage of two nights and three days, we saw few people. That is how it should be—that country needs to be used lightly to be understood and loved, to impart the feeling of freedom and the confidence and control that I shared with my friend on the trail.

Another day, I revisited well-established portions of the park on my own. Before walking the easy trails in Giant Forest, I read a few pages of John Muir's *Summering in the Sierra*. It recounted his adventures of 1875, when he studied the sequoia belt. Muir faced hard mountaineering before the winter storms, but was confident: "I will make a way, and love of King Sequoia will make all the labor light."[1] With Muir as guide and companion, the mind becomes receptive to more than eyes can see. At Giant Forest, I recalled Muir's description of "giants, grouped in pure temple groves."[2] These trees grow naturally in about seventy-five separate locations along the western Sierra slope, but Giant Forest is the largest of them all, enhanced by mild winters and gentle terrain. The giant sequoia (*Sequoiadendron giganteum*) and its relative the coastal redwood (*Sequoia sempervirens*) are the last surviving species of a large genus of ancient times.

The high country constitutes a vast region of unbroken wilderness, providing one of the outstanding experiences of the national park system. Motorists glimpse the wild country from overlooks, while the hiker journeys through canyons; across rivers, lakes, and meadows; and along the Sierra Crest. The principal footpath, the John Muir Trail, preserves the feeling of the original California.

Thus invigorated, I arrived in Berkeley to deliver the 1984 Horace M. Albright Lecture in Conservation. The lectureship was established in 1959 with a permanent endowment provided by friends and admirers, principally Laurance Rockefeller, who, for most of his life, was a friend and admirer of Albright.

I spoke to a large, receptive audience on the twentieth anniversary of the Wilderness Act, with due tribute to Albright, but I feel some other words I spoke then are fitting here:

> Freedom of expression is paramount in my life. I say that as a journalist, but I believe that free expression is the keystone of the health and efficiency of any institution or government or society. Diversity of opinion, even dissent, challenges an institution, or a political, social and economic system, to continually review and renew itself.
>
> As a journalist, I believe that truth-telling is essential to my profession. Truth-telling must and will prevail. "Knowledge will forever govern ignorance," wrote James Madison, "and a people who mean to be their own governors must arm themselves with the powers that knowledge gives." What greater goal could a journalist set for himself? What finer reputation could he earn than as one who arms the people with the power that knowledge gives?
>
> The same is true of anyone, for that matter, anyone with knowledge and position from which to communicate it. At times, to be sure, an open expression of ideas may seem foolhardy or risky. It endangers professional acceptance and advancement. But freedom of the individual, with the right of self-expression, is sacred. I consider my freedom as a need, like water or food, to sustain the spirit as well as the body; for real success or failure comes only from within and society cannot impress it from without.

To quote Joseph Wood Krutch: "Only the individualist succeeds, for only self-realization is success." Or, as a reader of *Field & Stream* wrote to me: "History books are records of events and the doings of individuals who didn't go with the flow." Let truth hang out and consequences follow.[3]

40

The Coalition Brings Its Experience to the Table

Over the years I observed that every new presidential administration, whether Democratic or Republican, arrives at the Department of the Interior with a clutch of political appointees eager to set a new tone in the management of the various agencies, including the National Park Service. In the case of the George W. Bush administration (2001–2008), the new team arrived fully charged, with a fundamental hostility toward career professionals running the parks. But that was the least of it.

It was part of a larger pattern designed to first degrade and then dissemble the established system of federal lands—embracing the national parks, national forests, national wildlife refuges, and areas administered by the Bureau of Land Management. As part of this process, the administration moved to "privatize" the management of public lands and "outsource" jobs to private contractors. Public Employees for Environmental Responsibility (PEER) warned that the administration was intent on replacing between one-quarter and one-half of all employees of natural resource agencies with private contractors, and opening a shift in management of national lands to states and localities.

I believe the rationale for keeping public lands public can be simply stated. Much of them are in the West. They are, in fact, the heart and body of the West, and maybe the soul too. Take away the public lands from the environs of Albuquerque, Boise, Denver, Phoenix, Salt Lake City, and Seattle, and they would be the most ordinary of places. Take away the public lands and there wouldn't be much to the economy either. Public lands are the last open spaces, last wilderness, last wildlife haven, last safeguard of watersheds. Without public lands the West would be an impoverished province. And the same can be said for the rest of the country as well.

The Bush-era war against the national parks derived from the Reaganomic formula of an earlier day that viewed all resources on 600 million acres of public lands as a source of potential private profit. The formula required the government to cut down the trees in national forests, increase the number of cattle grazing on public lands, allow the mining industry to pay little or nothing while polluting the environment, and provide more complex and undeserved subsidies to the oil industry.

The pattern of discrediting public lands was repeated across the board during the Bush administration. In numerous federal agencies, civil servants were frustrated and demoralized. They complained they weren't allowed to do their jobs because the White House operated on a very centralized basis, with bias to favor industry. Regulatory agencies—including the Interior and Labor Departments, the Environmental Protection Agency, the Food and Drug Administration, and the Consumer Product Safety Commission—were probably hit the hardest.

Hundreds of federally employed scientists, researchers, and agency attorneys drafted, studied, and restudied regulations that went nowhere. Political appointees, spread across the agencies, acted (a) for the interests of religious conservatives opposed to abortion and contraception, (b) for the interests of industry groups intent on blocking regulation, or (c) as messengers for an ideology that labeled government scientists as a class of liberals out to block the Bush agenda.

These appointees fell into repeated conflicts with agency career scientists. Deputy Assistant Secretary of the Interior Julie MacDonald, in particular, became legendary for meddling with Fish and Wildlife Service scientists whose research indicated a need for greater protection of endangered species. MacDonald was a civil engineer by training, with experience in California Republican politics. But her background, or lack of it, didn't stop her from challenging government wildlife experts and digging deep into their realm of expertise. According to an investigation by the Interior Department's inspector general that eventually led to her resignation in May 2007, MacDonald had been "heavily involved with editing, commenting on, and reshaping the Endangered Species Program's scientific reports from the field." That included demanding that scientists change their conclusions and reach ones more conducive to private industry.

Bush appointed industry people to key positions throughout the government. His secretary of the Interior, Gale Norton, had worked at Interior before, learning her lessons from James G. Watt, secretary under President Ronald Reagan. Like Reagan, Watt fought for free enterprise against the perceived evils of federalism and kept busy undermining and undoing established conservation programs until he became a liability and resigned.

Norton fit right in with Reagan and Watt. Later, as a lawyer in private practice, she represented polluters and corporate interests. During her five years in office as secretary, presumably protecting public lands and resources, she focused on opening the West to timber, oil, gas, and coal development, and to undoing generations of progress. When she left Interior she went to work for an oil company, Royal Dutch Shell.

Scandal, corruption, and misuse of authority plagued the Interior Department under Norton. Testimony at an oversight hearing of the House Committee on Natural Resources in mid-July 2008 revealed that four highly placed officials of the Bush administration had inappropriately influenced endangered species decision-making. "A disconcerting picture has emerged of officials working at the highest levels of the Interior Department continuing to tamper with the endangered species program, trumping science with politics. The practice is pervasive," said Representative Nick J. Rahall, a Democrat from West Virginia, the committee chairman. The hearing was held as a forum for the release of new findings in an ongoing Government Accountability Office (GAO) investigation into the well-publicized review of the politically tainted endangered species decisions made by MacDonald.

Perhaps the most serious case involved J. Steven Griles, the deputy secretary under Norton. He had earlier made a career as a coal, oil, and gas industry consultant and lobbyist; he, too, had worked at Interior under Watt. Griles resigned after an investigation revealed conflict of interest involving inappropriate communication with his former clients. He was also heavily implicated in the 2005 investigation of the notorious Jack Abramoff scandal—in which evidence showed Griles pledged to use his authority to block an Indian casino Abramoff was lobbying against. Two years later, in 2007, Griles was convicted of withholding information and sentenced to ten months in jail.

Curiously, Griles pleaded for leniency. He proposed to escape jail by substituting community service with the American Recreation Coalition (ARC), a Washington-based organization that promotes and lobbies for commercialization, privatization, and motorization of recreation in the national parks and other public lands. That did not fly.

If you were to ask at this point, "Why are the national parks administered in a department that apparently feasts on resource exploitation?" I would reply, "That is a very valid question, in search of discussion and solution, perhaps leading to reconstituting the National Park Service as an independent agency, like the Smithsonian Institution."

For the present, on with the steamy, stormy chapter of the Bush administration's war against the parks. Political appointees at Interior tried to sell off low-visitation parks to the highest bidder; they invited and promoted corporate branding of parks, and promoted outsourcing of maintenance, resource management, and science positions. They forced parks to reduce visitor center hours, curtail or eliminate interpretive programs, cut back on resource protection patrols, postpone cyclic maintenance programs, and leave key positions vacant. During the 2004 election campaign they claimed falsely to be spending more money per park, per visitor, and per employee than ever. And they silenced, or took punitive action against, career employees who said this was not so, and who attempted to "tell it like it is."

The principal political operative on park issues was Paul Hoffman, a former executive director of the Cody (Wyoming) Chamber of Commerce and a former Dick Cheney aide. As deputy assistant secretary, Hoffman, despite a total lack of scientific training, regularly overruled agency scientists on issues ranging from the effect of power plants on park visibility to whether cutthroat trout should be protected under the Endangered Species Act.

Besides the questionable management of park policies, Hoffman played a part in the dismissal of Teresa Chambers, chief of the U.S. Park Police (which operates in national park units in and around Washington, D.C.) for disclosing critical staff shortages in a *Washington Post* interview.

Chief Chambers brought suit for reinstatement, while she got a new job as police chief in Riverdale, Maryland. For five years she met with roadblocks and stubbornness from the Interior Department, rather than a reasoned settlement for injustice experienced. In November 2008 she dispatched an email report to her supporters:

> Our case still marches on. You will recall that we finally had an opportunity in November 2007 to present oral arguments to the U.S. Court of Appeals for the Federal Circuit. In February 2008, the appeals court determined that the Merit Systems Protection Board (MSPB) "applied an incorrect standard" in its failure to provide whistleblower protection status in my case and sent the case back to the MSPB to correct its failure. This marked only the third time out of 186 whistleblower cases between October 1994 and February 2008 in which the court ruled in favor of the whistleblower. A well written review of this ruling was published in the "Legal Times" and is linked on my husband's website, www.HonestChief.com.[1]

She and her husband waited patiently. In June 2009 the U.S. Court of Appeals unanimously upheld Chief Chambers's claim that records exonerating her may have been illegally destroyed by Bush administration officials, thus moving her one step closer toward getting her job back. But this was not the end. Interior Department officials from the administration of President Barack Obama fiddled and fumbled, supporting the position of the Bush administration, and the case went on. Finally, on January 11, 2011, a federal appeals board issued a definitive ruling restoring Chambers as chief of the Park Police. The Merit Systems Protection Board (MSPB), which hears civil service appeals, found that the evidence against Chief Chambers was weak and the motive of political appointees to retaliate was high. The MSPB ruled that no further proceedings were necessary and issued the following directive:

> Accordingly, we ORDER the agency to cancel the appellant's December 5, 2003, placement on administrative leave, cancel the appellant's July 10, 2004, removal, and restore her effective July 10, 2004. . . . The agency must complete this action no later than 20 days after the date of this decision.

"This is a wonderful ruling, not only for Chief Chambers but for thousands who believe that honesty is part of public service," stated Paula Dinerstein, counsel for PEER, who argued the appeals for Chief Chambers. "The wheels of justice turn slowly but eventually they do turn."

The director of the National Park Service during the dismissal of Chief Chambers was Fran Mainella, who came in with the Bush administration. Although a political appointee, she was a park professional, director of Florida state parks under Governor Jeb Bush, the president's brother. She wanted to do better in Washington, but became a messenger in the administration's war against the parks.

During Bush's first term, Hoffman was the principal political operative on park issues. But he ultimately overreached himself for his undoing. It happened that in 2001 the Park Service had adopted a mission statement declaring that "when there is a conflict between conserving resources and values and providing for enjoyment of them, conservation is to be predominant." But Hoffman secretly prepared a draft statement for a new mission statement that would subordinate the parks' conservation mission to "enjoyment" by the public—and clearly would open the way to hunting, snowmobiling, and other motorized intrusions. Hoffman learned the hard way that it's hard to keep a secret in government. The draft was leaked and sparked such a furor and firestorm of criticism that it was killed, and Hoffman was moved to a harmless administrative slot.

The whistle was blown on Hoffman in 2003 by three former National Park Service personnel—Mike Finley, Richard (Rick) Smith, and Bill Wade—at a press conference at the National Press Club in Washington. "The proposed changes," they warned subsequently, "are a drastic and dangerous departure from a longstanding national consensus. They are not driven by law, by any conservation need, or by any failure of practical application. . . . However, we are also concerned with opening the parks' management decision-making processes to disproportionate influence by special interest groups and local and individual state entities. These are national, not local, parks."

The three were surprised to be contacted by additional retirees who asked if they could join efforts to defend the parks and the programs of the National Park Service. Smith later explained, "They didn't like the fact that the whine of snowmobiles shattered the natural quiet of Yellowstone, disturbed park wildlife, polluted the air, and posed a threat to employee and visitor health and safety. They didn't see how programs like outsourcing maintenance, resource management, or science positions contributed to the effective management of parks. They felt that park visitors were not being adequately served."

That was the beginning of the Coalition of National Park Service Retirees, an organization that brought a strong and significant new dimension to the discussion and debate of national park issues. The coalition introduced the "voices of experience" to the table, with access to its own built-in network of current employees. Smith said:

> Most former employees care deeply about the traditions of the NPS and believe strongly in the management policies that have evolved over 92 years. They think the Organic Act is something to be respected, not an Act to be ignored when convenient to do so. They take seriously former Director Albright's challenge to not let the National Park Service become "just another government bureau." They don't like it when the Service's senior managers are ignored or bullied. They honestly believe that NPS decision-making should be transparent and based on the soundest science available.

In due course, in Bush's second term, Norton was replaced by Dirk Kempthorne, a former Idaho governor and U.S. senator, and Mainella was replaced by Mary Bomar, a career employee of the National Park Service. They softened the tone of their predecessors without deviating from the hard line. They both talked of "reinventing" national parks to attract more visitors, without reference to protection and preservation.

In its closing months, the Bush administration pulled out the stops. The president's men and women scrambled to change rules and regulations on the environment, civil liberties, and abortion rights, among others. Few changes were for the common good; they were more like a wrecking ball. It could have taken months, or years, to identify and then undo the damage.[2]

With the election the clock ran out and with it the wrecking ball. But the truth is that national parks suffered more during the Bush administration than during any other in my memory.

The establishment and growth of the Coalition of National Park Service Retirees gives me hope for the future. These people have not only worked in national parks but care about them, and are well worth listening to. One has to thank them for not just going quietly into retirement, which would be the easy thing to do.

This leads me to recall Bill Wade's review of the first edition of my book *Regreening the National Parks* (1991), published in *Ranger*, the journal of the Association of National Park Rangers. Wade wrote:

> Frome calls not only for a "regreening of the national parks," but perhaps more importantly, for a "regreening of the blood of those who manage and work in them." . . . *Regreening the National Parks* helped renew my sense of purpose and repaint my picture of what is right. More of us need to spend more time focusing on these concepts if we're really serious about our jobs.

I hope indeed that my work has helped to "regreen the blood." I take Wade's words as a compliment, considering he came up through the ranger ranks to direct the National Park Service training center at Harpers Ferry, West Virginia, and then was superintendent of Shenandoah National Park in Virginia.

Rick Smith of the Coalition of National Park Service Retirees reinforced this point, recalling in correspondence that over time I frequently mentioned the need for park managers to have a good moral compass and a strong set of ethics. Smith said:

> I fear these qualities are sadly lacking in the current crop, not only in Yellowstone, but elsewhere. Here in Yellowstone, it's hard to believe that snowmobiles still shatter the natural quiet of the park, harass park wildlife and significantly degrade the air quality in the winter. It's difficult to imagine that the park just caved in to the boosters in Cody and the politicians in Wyoming and decided to keep Sylvan Pass open for the pitifully small number of visitors that ride their snowmachines through the East Entrance.

The coalition, up to this writing, has proven itself as an expert witness willing to take on tough issues. These include opposition to a controversial rule issued by the Bush administration in the twilight of its days to allow visitors to carry loaded firearms in the parks. It was plainly a sop to the National Rifle Association (NRA) and the gun lobby. The coalition fought against the new rule, along with many other groups, saying it would only lead to increased "impulse shooting" of wildlife and risk the safety of visitors.

Doug Morris, who served for forty years as a ranger, manager, and superintendent of national parks from Alaska to North Carolina, said he never responded to a crime that would have been prevented had a visitor been carrying a concealed weapon. Nor did he hear complaints from gun owners about the rule requiring them to unload and lock away firearms while in national parks. But Morris did see cases where visitors shot wildlife or fired wildly into the night in crowded campgrounds.

Bill Wade was widely quoted with his statement: "This regulation will put visitors, employees and precious resources of the National Park System at risk. We will do everything possible to overturn it and return to a common-sense approach to guns in national parks that has been working for decades." Alas, the NRA had its way early in 2009 when Congress voted to authorize guns in national parks as an amendment to legislation dealing with credit cards. President Obama could have vetoed the bill and sent it back to Congress, but he signed it.

In September 2004 the coalition released "A Call to Action: Saving Our National Park System," a sixty-five-page document centered on this theme:

> Long-term strategic thinking is nearly impossible when the National Park Service is regularly whipsawed every election cycle by competing political ideologies. Management systems based on well-considered priorities are ineffective and failing. Federal budget cycles respond only to short-term considerations, and out-year planning is a constantly shifting political target for all levels of our governance system. "Park-barrel" decisions in Congress often lead to funding allocations that do not come even close to reflecting the highest priorities of the NPS, often leaving the real needs deferred or unfounded. Strategic plans become political statements for whatever party is in office rather than effective tools for agency governance of our precious heritage resources. Non-partisan politics that consistently characterized the early traditions upon which our national park system was founded are being replaced by divisive political ideologies that inject shameful political partisanship down to the lowest level of staffing and the most routine park management issues, destroying agency discretion and marginalizing the nation's best career professionals.[3]

41

Artists and Photographers Direct Us to a Sense of Place and of Spirit

While living in the Washington area, I found myself going to New York often, usually by plane, to interview someone of note, or confer with an editor or serve on a committee, and sometimes just for lunch.

That was how I met Alan Gussow in 1972, when we were on a committee together. I didn't know of him, as I might have, but learned that he was both an artist who painted from the landscape and a conservationist who tried to save parts of the world that were still natural. We communicated well during the course of the meetings, sharing common views and goals. In this respect he reminded me of my late friend Devereux Butcher, who, when he left his position with the National Parks Association, traveled to the parks and painted pictures of them. In the preface to the fifth edition of *Exploring Our National Parks and Monuments*, published in 1956, Butcher wrote, "Like literature, music and art in their highest forms, they [national parks] contribute to our spiritual well being, and vigilance to preserve them for that purpose."[1]

That is what I believe as well. Thus it was hardly a surprise to learn that in the fall of 1969 David Brower had invited Alan Gussow to speak on "The Artist and a Sense of Place" at a conference conducted by the John Muir Institute in Aspen, Colorado. As Brower explained, Gussow

> joined those trying to save the Storm King Mountain from Consolidated Edison's plan for meeting the demand for a never-ending increase in electrical energy. His defense of one of the places most important to him, the Hudson River Valley, was one of the most eloquent statements heard in the controversy. The

> eloquence of his brush was already proven. The show he put together for the Peridot Gallery in January 1968, "The American Landscape, A Living Tradition," led to the Smithsonian Institution's circulating the entire exhibit for a year. There was a book in what he had assembled, augmented by what he could say so well.

Gussow believed the artist's duty was to influence people to examine the world around them and to speak for change where need be. His activities in the Hudson River Valley led him to an appreciation of the national parks and involvement with them. In 1968 he served as the first artist-in-residence at a national park, Cape Cod National Seashore, and then as a consultant on the arts to the Department of the Interior. He felt there should be artists in national parks, just as there are poets and writers in residence at universities. Bill Wade, then an instructor at the Horace M. Albright Training Center in Grand Canyon National Park, later wrote to me that Gussow's presentations in the 1970s were well received and praised "not, I think, because of the beautiful pictures that accompanied his dialogue but because his presentation made participants think differently about the resources of the parks."

Much the same could be said about photography in the parks. Yosemite, as an example, is ideally meant for picture taking, with its granite domes and spires rising above glacier-carved canyons; waterfalls of unusual height and beauty; and the peaks, lakes, and meadows of the high country. From the foothills to the alpine zone, more than 1,300 species of flowering plants and many species of trees await the photographer. Many of the unusual rock formations are as striking in black and white as in color. Ansel Adams, known for his photographic interpretations of natural scenes, demonstrated this for years at Yosemite. He trained many park visitors with his summer workshops and books like *Gentle Wilderness: The Sierra Nevada*, published by the Sierra Club (with photos by Adams and captions to them from the writings of John Muir).

While Ansel Adams specialized in black-and-white photography at Yosemite, Josef Muench pioneered in color landscape photography in the Southwest. The Bavarian-born Muench for many years traveled across the region, photographing known and unknown places and people for the periodical *Arizona Highways* and now preserved in the book *The Southwest of Josef Muench: From the Pages of Arizona Highways* (1974).

Muench often was accompanied by his son David, who went on to a career of his own as a notable photographer of nature. In 1990 I was invited to collaborate with him in preparation of *Our Great Treasures: America's National Parks*, published by Rand McNally (1992). At that time David Muench explained that over the years he learned that the secret behind landscape photography is being in the right place at the right time, but, even more, knowing in your mind that the exact moment in space and time has been reached to record the perfect image of the subject at hand.

Much the same could be said for art in the parks. That is what Gussow and I have tried to do—get people to think differently about our national parks. When his book *A Sense of Place: The Artist and the American Land* was published in 1972, he gave me an autographed copy that I consumed with gusto. It covers four generations of works by American landscape artists, complete with sixty-three color plates, and excerpts from the artists' writings. It includes paintings of the nineteenth-century Hudson River School, America's first national school of landscape painting, and of its founder, the British-born Thomas Cole (1801–1848), who produced work that was visionary, panoramic, and serene, with mountains veiled in mist. Hudson River School works of art are now found all around America, and in other countries as well. It was the virgin character of America's wilderness, in contrast to the more civilized appearance of the Old World, that Cole and others sought to depict. From the 1820s through the 1880s they evoked symbolic images of the wild even as America hastened to expand westward to pursue its "manifest destiny," ever yet in our age unmet, unfulfilled.

As the country moved west, painters ventured into the wilderness to record its grandeur and to generate a national conscience to preserve and protect. The great canvasses of Thomas Moran (1837–1926), artist of the 1871 Hayden Geological Survey, strongly influenced establishment of Yellowstone as a national park. Later he went to the Grand Canyon and in 1874 completed *The Chasm of the Colorado*, destined to hang in the lobby of the U.S. Senate in Washington. Meanwhile, the German-born Albert Bierstadt (1830–1902) went west from New England to produce the massive landscape *The Rocky Mountains, Lander's Peak*, 1863, that hangs prominently at the Metropolitan Museum of Art in New York.

Bierstadt also went to California, the source of his renowned and dramatic paintings *In the Yosemite Valley*, 1866; *In the Mountains*, 1867; and *The Hetch-Hetchy Valley*, ca. 1874–1880.

George Catlin (1796–1872) began his professional life as a lawyer but gave it up to live with and interpret on canvas Native Americans in the beauty and variety of their native landscape. He was moved, in fact, to urge establishment of a great national park in the heart of the continent where American Indians, their lifestyle, and their lands would be safe. He gathered five hundred of his paintings, plus native artifacts, for his "Indian Gallery," which he showed in New York and other eastern cities.

Rockwell Kent (1882–1971) would become a preeminent American artist and illustrator, but in 1918 he was living, for more than half a year, with his nine-year-old son on a primitive island near Seward, Alaska; the island has since become part of Kenai Fjords National Park. Kent's "adventure of the spirit" became the basis of his book *Wilderness: A Journal of Quiet Adventure in Alaska*, published in 1920 to stellar reviews, and reissued in 1996. I learned, when my wife and I visited Kenai Fjords, that reissue was due in no small measure to the active interest of Doug Capra, a park naturalist, who in the foreword writes:

> Since its publication, *Wilderness* has stirred the imagination of countless artists, writers and adventurers and influenced more than one of them to venture to Alaska. It is a book about art and life, about alienation and integration, about the inner life, the spiritual life, the simple life, and about growing old gracefully, without losing one's childhood ideals. In *Wilderness*, Kent confronts the emptiness and loneliness of the abyss and fills it with the richness and wealth of his soul.[2]

And that, in extension, one might say, is the role national parks fill in American life. As for art and nature, Gussow sums up the collection of artistic treasures in *A Sense of Place*:

> These paintings tell us something very old and therefore very new about knowledge. They tell us that the whole world is still much lovelier than can be explained by everything that's in it, and that however much we learn about the pieces, the whole is still a sum beyond their totaling. These paintings remind us

> of the human event. In a time that too often confuses data with experience, the sun is still a yellow ball pressing down on us from the sky at midday. Its color and its temperature affect us. It is this hot and vivid sun to which these artists respond.
>
> What is under stress today is the very definition of the natural. What is natural to many people, whether we like it or not, is what they have grown up expecting, and in our man-made world nature is increasingly unexpected. Those of us who believe that natural places are necessary for the salvation not only of the human race, but also of the human spirit, must insist on pointing up, over and over again, what it is we risk losing.
>
> Nineteenth-century painters went out into the wilderness to bring back reports about a land we did not know; painters now report a land we risk forgetting. Their gentle paintings direct us earthward; they remind us of seasons, of time of day, of processes outside human factors. They urge upon us a balance. They do not suggest a return to rustic simplicity wholly inappropriate to our times. They are not retreats. These paintings put a value on certain qualities in the environment. Acts of salvage in a desperate time, they can become a model against which we measure our success or failure as restorers of the land.[3]

It grieves me to record that Alan Gussow died in 1997. But I can express the hope that his words and work direct us earthward and they encourage and help us to serve as restorers and protectors of land, and of spirit.

42

Brower, Without Fear or Favor

David Brower was plainly weak when he came to speak at the Northwest Wilderness Conference in Seattle in April 2000, barely eight months before he died. He walked with a cane and needed help getting up to the platform. He looked his years, eighty-eight, but when he spoke he was as strong as ever—strong in commitment, powerful in inspiration. When he died that November, the *San Francisco Chronicle* described him as "the most influential figure in the American environmental movement since John Muir and Theodore Roosevelt . . . utterly uncompromising on stewardship of the planet."[1]

I agree. Without fear or favor, Brower championed naturalness and integrity in the national parks and in American life. His genius lay in his clarity of purpose: always creative and daring, no mountain too high to climb, no battle of principle too tough to fight.

I was with Brower on the platform at the conference in Seattle. He and I comprised one-half—call it the male half—of a panel of seniors, or old-timers, who were assigned to look back and look ahead. I knew both women well, and respected and admired them.

Polly Dyer was my age, born in 1920, energetic, tireless, active for more than half a century in every major campaign to protect wilderness in Washington State, including the establishment of North Cascades National Park. When she celebrated her ninetieth birthday, more than two hundred friends and admirers came to pay her tribute.

Celia Hunter was also of my vintage. She served during World War II as an air transport pilot, shuttling fighter planes across the continent. Following the war, she and her partner, Virginia Hill Wood, headed north in two beat-up warplanes. Ultimately, they staked a claim on a blueberry-covered ridge just outside what was then called Mount McKinley National Park, dreaming of establishing a wilderness retreat where they could provide comfortable accommodations without destroying the beauty they beheld. They became as self-reliant as any men in the bush, building the complex of tent cabins and rustic chalets called Camp Denali, with a majestic view of the tallest mountain on this continent. They ran it for twenty-five years before turning it over to younger hands. Hunter, whom I always found handsome and vigorous, became a humanitarian and one of Alaska's pioneer environmentalists.

Brower was a true believer, but several levels above all the rest. In addition to being the focus of John McPhee's *Encounters with the Archdruid* (1971), Brower was responsible for another epochal literary work. In December 1967, after reading an article in *New Scientist* by Paul Ehrlich, a Stanford University biology professor, Brower contacted Ehrlich with a strong recommendation that he make a book out of the article. That was the beginning of *The Population Bomb* (1968), which ultimately woke up America and the world to the presence of too many of us on this planet.

Brower was tall, about six-foot-three, erect and athletic from his days in mountain climbing, and prematurely white-haired. He had attended the University of California for four years without graduating, and worked as an editor for the University of California Press and as a PR person for the concessionaire in Yosemite National Park.

In 1957 he helped a handful of Washington State activists organize the North Cascades Conservation Council. "In the early 1960s, as it grew obvious we locals had to "go national," Harvey Manning later recorded, "Dave's leadership became paramount. He knew all buttons of all the players in the national game. He pushed them. And thus, the North Cascades National Park was created in 1968."

Brower was gifted artistically and might have been successful as an author, editor, or filmmaker. To advance the North Cascades cause he produced and personally photographed a beautiful documentary film, *The Wilderness Alps of Stehekin,* with voices of the Vienna Boys Choir. He launched the distinctive Sierra Club Format Books with *This Is the American Earth* (1960), written by Nancy Newhall and illustrated by Ansel Adams. Commercial publishers later copied the style, but they produced mere coffee-table books.

While doing such creative things with one hand, with the other hand Brower hired Gordon Robinson as the Sierra Club's first professional forester in 1966. Until then the irate public had felt ill-equipped and inadequate to debate the Forest Service and timber industry over technical issues of "allowable cut effects" and "mean annual increments." Robinson was able to address these issues in a professional manner. He was a graduate forester (from the University of California, Berkeley) who had worked for many years managing large blocks of commercial timberland for the Southern Pacific Railroad, the largest private landholder in California.

When he joined the Sierra Club staff, Robinson confirmed suspicions that something terribly wrong was taking place. Now at last the club and environmental movement could decode the jargon, penetrate the curtain of professional expertise, and even the scales in debate. Robinson testified before Congress, served as a witness for local groups, and in general revived idealism and courage in forestry—the tradition of Gifford Pinchot, Aldo Leopold, and Bob Marshall. In 1988, a few years before his death, he published *The Forest and the Trees*, defining the principles of true multiple-use forestry.[2] His work is still valid today.

Brower also hired Brock Evans, an idealistic young lawyer who worked in Seattle on major Northwest issues, and later in Washington, D.C. He later told his story as follows:

> I first came on the scene in mid-1967. From David Brower, who had hired me just a few months earlier, I learned that my "territory" was, literally, everything from the North Pole to San Francisco, and as far east as Yellowstone. Northwest North America: from Alaska to the North Cascades to the Oregon Cascades, from the Sawtooths to the Flathead to the Wind Rivers, and every place in between. Since I was the only full-time paid person north of San Francisco, I felt responsible for the fate of Hells Canyon—at least accountable for putting together a campaign that we all could rally around and have a chance of success. I hadn't yet met anyone from the just-formed HCPC (Hells Canyon Preservation Council), so except for a small and courageous nucleus of local folks, I felt quite alone.
>
> I had another concern, more personal. I loved the place. From the first time I had journeyed through the Canyon a few weeks before, enchanted by its majesty and beauty, it was as if some old lost chord had been plucked inside. My heart sang

> to a new kind of music I hadn't even known was there. I loved Hells Canyon, and vowed to give everything in my power to try to save it.[3]

Brower came across at times as reserved and distrusting, perhaps aloof or arrogant. He had his detractors and critics, from the directors of the Sierra Club to academic historians who discounted his "high-handed style." But the style in question was "high-handed" principally to those who objected to his dauntless efforts on behalf of wild America. Martin Litton, his close ally, saw Brower as so consumed with the next crusade that he didn't have time for small talk with mere mortals.

In 1969 Brower resigned and established Friends of the Earth, where similar problems arose, leading to another departure. From there he went to Earth Island Institute, which is still going strong. True believers are that way. They believe in better days and better ways, in transformation of the individual and of society. "Tomorrow may be the day of judgment," the true believer might concede, then adding, "If it is, we shall gladly give up working for a better future, but not before."

Prior to his death in 2000, Brower wrote the foreword to a version of Manning's *Wilderness Alps: Conservation and Conflict in Washington's North Cascades*. It includes these immortal lines: "Dream a bit about what will happen if instead of trashing still more, we determine that it's healing time on Earth, and we're not going to make birdwatchers irate any more."[4]

43

Failing to Safeguard the Sacred, Ancient, and Fragile

Of the various cultural resources presumed protected in our national parks, in a 2008 study of the National Academy of Public Administration, archeology in particular was found seriously wanting, and unprotected. To cite the study: Archeology needs "more focus on protecting sites." Of the 84 million acres managed by the National Park Service, only two percent has been surveyed for archaeological resources, while less than half of the identified sites are in good condition. The Washington office archeology staff consists of only five professional archeologists.

In 2005 the National Parks Conservation Association asked me to visit Petroglyph National Monument in Albuquerque, New Mexico, which was then embroiled in the age-old controversy between preservation and commercial development. I found the monument embraced more than twenty thousand petroglyphs, or etchings in volcanic rock, and was a sacred religious site still in use by the area's Pueblo tribes.

When Congress established the national monument in 1990, in order to "protect the cultural and natural resources of the area from urbanization and vandalism," it rescued parts of the area from decades of abuse as a dumping ground and shooting gallery, with the petroglyphs frequently serving as targets and with graffiti spray-painted on ancient images. This congressional action came in response to a campaign initiated by Friends of the Albuquerque Petroglyphs, organized in 1986 and spearheaded by Isaac "Ike" Eastvold, who had chosen to devote much of his life to the protection and interpretation of archeological sites in and around Albuquerque. Eastvold would later be honored for his efforts by the National Geographic Society and National Parks Conservation Association.

Commercial development interests lobbied for a main-stem highway through the landscape presumably protected by the monument. The New Mexico congressional delegation introduced bills on their behalf to delete eight and a half acres from the monument. Paseo del Norte extension, the proposed six-lane highway, would not only accelerate urbanization but violate the integrity of the monument and, by extension, the national park system.

En route to Albuquerque, I thought of an earlier visit I made to Chaco Culture National Historical Park, located in the arid and sparsely populated Four Corners region of New Mexico. It embraces the densest and most exceptional concentration of pueblos in the American Southwest, a sweeping collection of ancient ruins, one of the United States' most important pre-Columbian cultural and historical areas. I learned there that many Chacoan buildings may have been aligned to capture the solar and lunar cycles, requiring generations of astronomical observations and centuries of skillfully coordinated construction. The sites are considered sacred ancestral homelands by the Hopi and Pueblo people, composing a UNESCO World Heritage Site.

The same might have been true of the petroglyphs at Albuquerque, a repository of archeology and aboriginal art. While touring the site, I observed that some panels showed distinct figures of humans and animals, done in red, yellow, orange, and black, with symmetrical, stylized forms. Others were faded and dim, beyond recognition. Here was the opportunity to reconstruct prehistory, to explore the evolution of American primitive art.

As a progressive candidate for mayor in 1997, Jim Baca opposed the road through the monument and was elected. He advocated a regional growth plan to prevent urban sprawl. "Extending Paseo through the petroglyphs won't solve any traffic problems," he declared, "and would create a bad precedent for deleting lands from national parks for development purposes."

Times changed. The *Albuquerque Journal*, the largest newspaper in New Mexico, and the evening *Albuquerque Tribune* (now defunct) beat the drum for the roads through the monument. The *Journal* editorial page editor, Bill Hume, used his editorial and op-ed pages to attack and belittle Eastvold.

The All Indian Pueblo Council, representing the nineteen New Mexico pueblo tribes, whose ancestors carved the petroglyphs, opposed the Paseo extension on religious grounds. So did the Navajo Nation, the National Congress of American Indians, and thousands of individuals and organizations.

Friends of the Albuquerque Petroglyphs initiated the publication of a book, *Voices from a Sacred Place: In Defense of Petroglyph National Monument*, edited by Verne Huser. When the book appeared in March 1998, it featured contributions by forty writers located in different parts of the country, yet united in this common cause.

In an essay titled "The Center of the Spiral," Stephen Trimble wrote:

> Build a road through these sacred artscapes? Pueblo Indian people see this as a violation of holy ground. How can non-Indians understand this? Is this land sacred to all? Can we base public policy on the fundamental belief of Indian people—that everything is sacred, that everything is connected?
>
> David Brower gave us an answer thirty years ago when he opposed dams in the Grand Canyon: we do not flood the Sistine chapel to get a closer look at the ceiling.
>
> We do not drive our vehicles into the naves of churches, the arks of synagogues, under the domes of mosques, or down the ladders of kivas. We do not talk on our cell phones while we pray.[1]

I also contributed an essay to *Voices from a Sacred Place*. I wrote that it saddened me, while in Albuquerque, to read a statement attributed to the monument superintendent, Judith Cordova: "You can't please everyone. We are not a rural area. We are an urban area." That did not ring true, coming from the guardian of such a choice living cathedral. Consequently, on returning home, I wrote for clarification to John Cook, the Southwest regional director of the National Park Service. I hoped he would agree that personnel of his agency are not mandated to please everyone, but to do their best to protect the treasures in their trust. It grieved me when the letter I received from Mr. Cook reiterated the same old political pap about "conflicting public ideas" and "appropriate balance" between preservation and use.

In October 2003 voters soundly rejected a bond issue to partially fund construction of the road extension. J. J. Brody, emeritus professor of art and art history at the University of New Mexico, cheered defenders of the monument by declaring in March 2004: "The proposed road will seriously damage the physical integrity of the Monument, generate sound and air pollution that will degrade it, and stimulate more urban sprawl whose costs will be borne by all Albuquerque taxpayers."[2]

But pro-growth proponents, politicians, and developers kept coming on and the campaign gained steam. The issue here was not unlike that in Memphis, Tennessee, during the 1960s and 1970s when a band of energetic citizens fought to protect Overton Park, one of the finest urban forests in the world, from proposed construction of a highway through the middle of it.

Ultimately, Overton was spared; it still enhances the landscape and quality of life of Memphis. The same, alas, cannot be said for Albuquerque. Senator Pete Domenici's bill to delete eight acres from the monument to provide for the road was never discussed in Congress; nor was it voted on as such. Domenici attached it as a rider to a several-billion-dollar bill to fund operations in Bosnia, leaving the middle of the monument open to road construction. Senators do such things. And thus the Paseo del Norte extension was built, violating the seventeen-mile-long monument, opening the way to runaway growth and massive traffic.

At one point I recalled a visit I had made many years before to the cave at Altamira, near Santander in northern Spain, renowned for its Upper Paleolithic art, including drawings and polychrome rock paintings of wild mammals and human hands. The study of Altamira changed forever the perception of prehistoric human beings and it now is a UNESCO World Heritage Site. Because the paintings were being damaged by the carbon dioxide in the breath of the large number of visitors, Altamira was completely closed to the public, with a replica, including all of the art, built nearby.

That was how Spain treated its ancient cultural treasure. As to the story of the Albuquerque petroglyphs, there is this to add from a 2009 letter to Secretary of the Interior Ken Salazar from Dr. Harvard Ayers, Professor Emeritus of Anthropology at Appalachian State University, who had served on the Petroglyphs National Monument Advisory Management Commission in the mid-1990s:

> When the PNM Advisory Management Commission was determining our recommendations, we were quite concerned with the broad "recreation" focus of the park being proposed by

> NPS planners. We voted unanimously against that focus in recommending against allowing horseback riding and mountain biking on the mesa-top part of the Monument that is an integral part of this sacred landscape to the Pueblo people. All nineteen Pueblo nations also signed a letter to the Park Service with the same recommendations. . . .
>
> Over the 17 years since the Monument was established, I have visited once or more every year and have witnessed the results of the lax management practices. No attempt that I am aware of has been made to manage visitors in any of the culturally sensitive areas of the Monument. . . .
>
> In my estimation, the park is being "loved to death." Of course, this is a suburban park. And some accommodation of the local neighborhood citizens is appropriate. But totally unsupervised and complete access is leading to severe deterioration of the cultural resource. The sacredness of the park does not seem to be a concern of the management plan, and obviously the atmosphere of this special place has suffered considerably. If these trends are allowed to continue, except for the obvious threat of encroachment by the surrounding subdivisions, the Monument may have been better off not being so designated.

Shortly after writing this letter, Dr. Ayers related to me a telling incident: "One thing I remember very clearly is that even after this unanimous recommendation Superintendent Judith Cordova told the commission, 'You are only advising. You don't have any power.'"[3] This sort of episode is not the exception, but more the rule in dealing with particular archeological treasures.

Here now, another painful case history: Paul Berkowitz, a former ranger and veteran special investigator on the National Park Service payroll, told all in his 2011 book *The Case of the Indian Trader: Billy Malone and the National Park Service Investigation at Hubbell Trading Post*. He recounted the story of Billy Gene Malone, who lived most of his life on the Navajo Reservation in Arizona working as an Indian trader, and then came to operate historic Hubbell Trading Post at Ganado.

John Lorenzo Hubbell opened this trading post on the Navajo Reservation in 1878. He became a respected personality who advised the Navajo to use natural dyes in rugs and blankets and made Navajo weaving an art form that helped them recover from death and devastation. In recent times the trading post was purchased by the National

Park Service from the Hubbell family and in 1967 it was added to the national park system as a national historic site, one that would be run as an authentic trading post, with an authentic Indian trader. That was Malone.

Although he worked closely with the National Park Service at the Hubbell Trading Post, Malone was an employee of the Southwest Parks and Monuments Association, soon to become the Western National Parks Association (WNPA). The new management team of WNPA became convinced that Malone was guilty of fraud and urged the National Park Service to fire him and open a criminal investigation. In 2004, the Park Service thus targeted Malone, alleging a long list of crimes, misdemeanors, and mismanagement. In 2005, Berkowitz was assigned to take over the year-and-a-half-old case. His investigation uncovered serious problems with the original allegations, raising questions about the integrity of supervisors and colleagues, as well as high-level Park Service managers.

Wary of how his findings might be handled, Berkowitz presented his report not to the Park Service but to the Interior Department's Office of Inspector General (OIG). In the end, he walked away from his career and retired. National Park Service officials since then have either ignored requests to discuss the Hubbell episode or cite ongoing legal matters that prevent them from speaking. But then, while the trading post looks authentic from the outside, on the inside it is run more like a park concession, a bookstore, or a gift shop than an actual live trading post.

In his book, Berkowitz recognizes that many in the Park Service ranks are honest, hard-working, and dedicated. But he is relentless in his criticism of what he sees as corruption, cronyism, nepotism, and lack of respect for law and policy within the leadership:

> With its appealing mission and characteristically idyllic and isolated work environments, a very distinct culture and psychology have evolved within the NPS. NPS employees tend to think of themselves as separate and apart, special and elite within the government. They are not merely loyal civil servants in the employ of a unique federal agency. They are lucky and privileged members of a workforce internally characterized as a family, and its good and loyal members demonstrate in both their work and their personal lives that they have green blood

> flowing through their veins and are "green and gray" through and through. Employees are indoctrinated into this culture and encouraged through a system of very real social and professional incentives and sanctions to commit themselves fully to the agency whose mission and traditions they embrace.[4]

Tough talk, but I believe he's right. The National Park Service tends to ignore, or to punish, the expression of divergent points of view within the ranks. In the face of criticism, it tends to circle the wagons, hiding behind bland and deflecting platitudes, even when the criticism is well founded, and when looking for the lesson is clearly the better course.

44

Sometimes Rules and Regulations Are Bent and Broken

Despite efforts of many able people in the ranks, and supporters on the outside, the National Park Service, as an organization or institution, has become part of the problem instead of the solution. The way things work, regulations, rules, and laws are bent, broken, or ignored. The National Park Service should be a leading force for conservation and the land ethic, but it is not.

"The NPS has provided not a shred of environmental leadership," Frank Buono, who served in key positions for many years, wrote to me late in 2011. "One recent example—fostering gas-guzzling, carbon spewing ORV [off-road vehicle] use in Big Cypress [National Preserve, in Florida] and doing so by dishonest reviews that wave a magic wand and puff!—40,000 acres of roadless land are not wilderness eligible anymore. The agency is leaderless. NPS sails catch every passing breeze—and thus cannot steer a straight course. The concerns for childhood obesity, healthy lifestyles, dietary habits, cyber-connectivity, though worthy, all have become the agency mission. When, in fact, none is."

Buono's point is not new. In November 1996 I received a letter from Howard Chapman, retired regional director in the West (mentioned elsewhere in this book), which includes the following:

> The NPS complains bitterly about not having enough money to protect the parks. Yet each year they get more money than the previous year, while citizens hear about budget deficits and balancing the budget. The public gets less interpretation but the Service continues to send people to meetings and lay plans for expensive improvements that will also be costly to maintain.

> There is no strong hand today. More money and partnerships seem to be what consumes them instead of what is really important in the parks and what they will mean to future generations. Somehow the American public needs more than a cry for more money—the public needs to know what is REALLY at stake!

On the whole, the National Park Service culture is resistant to change, and to serious citizen input. For example, the bottled water fiasco began in 2010 when Grand Canyon National Park announced it would prohibit vendors from selling water in the park; instead, it would offer stations for reusable bottles. That was to the good, considering that bottles have been the single biggest source of trash found inside the canyon. Zion National Park, in Utah, had already instituted a similar ban on bottles in 2008, eliminating sixty thousand plastic bottles from the park in its first year; that move, understandably, was met with public appreciation and praise.

The plan to ban the bottles in the Grand Canyon was on track to take effect January 1, 2011. But Coca-Cola complained to the National Park Foundation, the nonprofit agency that channels financial support to parks from private companies and donors. In short order, National Park Service director Jon Jarvis intervened to block the ban and place it on hold. Since Coca-Cola has donated $4 million to the foundation for various projects, including trail maintenance and tourism promotion, park supporters suspected a case of undue influence and, even worse, of creeping corporatization in the parks.

Extensive media coverage led to a national campaign with more than 92,000 people urging the National Park Service to proceed with its plans to ban bottles at Grand Canyon. The National Park Service reversed course. On December 14, 2011, Director Jarvis announced that, "in light of recent interest," he was backing down and allowing park superintendents to halt the sale of plastic water bottles. The truth was that the Park Service bent backward toward its commercial allies until a public uproar made it straighten up and do right.

Another case: In 2011, after dogged Freedom of Information Act requests that produced hundreds of pages, *Seattle Times* reporter Ron Judd recorded in detail how David Uberuaga, as superintendent of Mount Rainier National Park, said he did not consider it a conflict of interest when he sold his home in 2002 for more than three times its assessed value to one of the park's concessionaires.

Uberuaga had bought his house in Ashford in 1992 for $84,000, then he sold it in 2002 to Rainier Mountaineering's Peter Whittaker, the park's climbing concessionaire, for $425,000. The property's assessed value was $122,400. Because it was so far above value, a conventional loan was unavailable, so Uberuaga financed the deal and Whittaker paid him principle and seven percent interest over five years. Then Uberuaga was promoted to superintendent of Grand Canyon National Park.

Uberuaga never recused himself from park-business dealings with Whittaker and repeatedly offered only a vague description of the deal on his federal ethics-disclosure forms, according to records reviewed by the newspaper. Even when pressed for more details by a federal ethics officer, he failed to immediately reveal that the man paying an unusually high price for his home also owned the park's largest concessions contract.

In an interview, Uberuaga conceded the sale looked suspicious, but said it wasn't a kickback, and at one point he passed a polygraph examination about that. He also said he never intentionally withheld information.

Not only was the Interior Department's Office of Inspector General (OIG) troubled by the sheer price of the sale, but it also suggested a possible conflict in that Uberuaga oversaw park concessions. OIG investigators were also troubled that the sale was not financed in the typical way, through a bank. Rather, Uberuaga held the note on the sale.

A review of this episode strongly suggests that National Park Service rules are meant to be bent by officials in the inner circle of power. It raises a question of ethical standards during the tenure of Director Jon Jarvis. America's national parks deserve stewards whose reputations are not sullied, whose judgment isn't clouded, who have proven themselves to be the best, with unblemished records and the utmost confidence in them to do their jobs appropriately. This goes for Jarvis as well.

For many years, the Park Service as a whole has resisted and opposed implementation of the 1964 Wilderness Act. But it doesn't mean everybody feels that way. Some actually believe the agency has a missed responsibility here. On May 5, 2000, Richard W. Sellars, a Park Service historian and author of *Preserving Nature in the National Parks*, submitted a document titled "Comments to the National Park Service's National Wilderness Steering Committee."

The following paragraph may be considered as the synthesis of this document:

> A number of upper-level managers within the National Park Service appear to believe that wilderness management plans are not necessary. Yet, I feel strongly that the plans form a kind of contract with the public, with the National Park Service itself, and with a park's subsequent personnel, by stating the methods and means by which wilderness will be managed. Among other things, park wilderness management plans require: an organizational profile that clearly identifies responsibility (i.e., who is accountable) for wilderness management and preservation; clear minimum-requirement protocols; clear protocols for scientific research in wilderness areas; assurance of full integration of wilderness preservation into long-term and day-to-day park operations; and the clear identification of legal boundaries for wilderness. The plans thus provide a detailed blueprint by which responsible parties can be held accountable.[1]

Despite the evidence and suspicions of cronyism and corruption, my own personal experience has shown that many, many in National Park Service ranks are able, honest, committed, and dedicated to the public interest.

To illustrate the positive, in 2005, when my wife and I were in Little Rock, Arkansas, I was especially interested in seeing the new Little Rock Central High School National Historic Site. At the visitor center (converted from a Mobil Magnolia service station across the street from the school), we received a healthy history lesson, learning of how Central High was completed in 1927 with a twenty-one-acre campus and capacity for two thousand students. But it was built and open for white students only. That was how they did things in the South. Two years later, Paul Laurence Dunbar High School was built a few blocks away. It was plainly unequal in all respects: quality of construction, operational support, quality of supplies, qualifications of teachers.

The unequal treatment of races set the stage for the Civil Rights Movement of the 1950s and beyond, leading to the Supreme Court decision overturning the "separate but equal" doctrine and mandating desegregation "with all deliberate speed." In 1957 the school board actually planned for quiet compliance with the Court's ruling in *Brown v. Board of Education*, but Orval Faubus, the segregationist governor, proclaimed the cause of states' rights and himself as defender of white pride. He called the National Guard to "keep order" by blocking the entry of nine black students. With an angry prosegregation crowd on the school steps, the Little Rock scene was shown on TV screens and front pages across the country. President Dwight D. Eisenhower, believing deeply in the rule of law, sent troops to escort the nine students on September 25, 1957. Consequently, the students were assigned guards to walk them from class to class in the face of physical and verbal harassment. A white student later said, "It was the first time I'd ever gone to school with a Negro, and it didn't hurt a bit." Over the years the "Little Rock Nine" received numerous awards, including the Congressional Gold Medal. Each of them graduated from college and followed professional careers. A movement arose in Little Rock to tell their story and to make it the symbol of the end of racially segregated public education in the United States.

The Little Rock Central High Museum board, supported by state and local governments, sought to have the site added to the national park system, and broached the idea to two Arkansans in high places: President Bill Clinton and Senator Dale Bumpers. The effort was helped further when Don Castleberry, a Little Rock native, returned home and joined the board, following a thirty-two-year career in the national parks.

The Little Rock Central High School National Historic Site was established to preserve, protect, and interpret "its role in the integration of public schools and the development of the civil rights movement in the United States." A new $6 million visitor center was dedicated in 2007. All of the "Little Rock Nine" were in attendance and duly honored.

Young people today may not know, and older people may long have forgotten, but fifty years ago the nation watched transfixed as nine black students attempted to enter the previously all-white school. I doubt this episode is cited much in high school history classes today. Yet it should be marked and memorialized, as a living reminder of how history was made and how Little Rock came to symbolize the national commitment to eliminate separate and unequal systems of education for blacks and whites.

On another front, I well remember how in the 1980s the California condor, the largest North American land bird, was at the brink of extinction after its numbers plummeted due to poaching, lead poisoning, and habitat destruction. All looked lost in 1982 when the last twenty-two California condors were placed in a captive-breeding program. It became the most expensive species conservation project ever undertaken in the United States, but it worked. Numbers rose and condors were reintroduced into the wild.

During the condor's slow recovery, biologists at Pinnacles National Monument in central California celebrated the first condor egg laid by a mating pair inside the park boundaries in more than a century. "We are thrilled that after being involved with the Condor Recovery Program since 2003, the park has its first nest in over 100 years," said the park superintendent. As of December 2011, there were 390 condors known to be living, including 210 in the wild.

Another encouraging case: The Sea Turtle Science and Recovery Program at Padre Island National Seashore reported from the Texas Gulf Coast that fifty-plus hatchlings of the endangered Kemp's ridley sea turtle were released from their nest at the national seashore, and that a green sea turtle nest was found, bringing the total to five green sea turtle nests found in 2011, all at the national seashore. That is truly turning the tide, after long abuse of sea turtles and their habitat.

In contrast to condors and turtles, here is a new park mission, meant to conserve a distinct American culture dating to the early days of slavery in the South. Based on a congressional directive, the National Park Service in 2010 began studies of how to protect and interpret features of the Gullah/Geechee Cultural Heritage Corridor encompassing coastal communities from Wilmington, North Carolina, through South Carolina and Georgia, to Jacksonville, Florida. Along this corridor, historic plantation lands and communities have been inhabited by Gullah/Geechee people for more than three centuries. Here, enslaved Africans and their descendants worked the rice, cotton, and indigo plantations that characterized the South. It may be a different approach, but important places, things, and traditions should be recognized and shown.

These last citations show a very positive side of the National Park Service, free of visitor numbers. They demonstrate what park people can do and what our national parks should be.

That is all to the good. But then, alas, there is the other side. Before coming to Yellowstone, for instance, you think of this great park the way it is shown in the nature series on television. The Park Service tells you to unplug your ears and connect with nature—but when you arrive you find visitors concentrated in developed areas, with electric wires, phone lines, lots of automobiles, gas stations, hotels, commercial gift shops, and sewage treatment. These, in NPS parlance, are "sacrifice areas," otherwise known as popcorn playgrounds or tourist ghettos.

In 2013 Yellowstone park administrators said there would be no significant impact from the introduction of cell towers. To be sure, those cell towers would enable you to check your email, track your stocks, and feel the conveniences of home.

Perhaps park administrators might have chosen not to allow those towers in the first place. They might have determined this was a strictly commercial service using public resources and public land, and that the signals the towers emit can spill into and pollute hiking trails away from developed areas. They might have decided that since hotels in the park get along without television, they can make it without wireless Internet service. When people come to Yellowstone, it's one of the special times in their lives. They want to hear the splash of geysers and feel themselves in harmony with natural forces that over the centuries created the thermal features, peaks, and canyons. That is what they come here for, and not having that sound drowned out by somebody conversing via cell phone.

National parks are presumed preserved to reflect the original America. Many National Park Service personnel want it that way. They care deeply, feeling that their mission is to encourage us to embrace a lifestyle that treads lightly on the earth, and that doing so adds richness to all of our lives. They ought to be able to defend their park areas from overuse and misuse with a clear conscience. To deplete or degrade the visible physical resource does something to the invisible spirit of place as well. Native Americans have that ancestral sense, honoring the earth and life as divine gifts, and modern Americans should learn from them. In the 1925 film *The Vanishing American*, Zane Grey's hero, Nophaie, most loved to be alone, out in the desert, "listening to the real sounds of the open and to the whispering of his soul."

In short, instead of treating a national park like any other place, the park professionals ought to say, "If you can't do without your cell phone or laptop or tablet, don't come here."

In 2013 I received an email from a Park Service retiree asking a question, "Will they ever learn?" Then he went on, "This is why

Yellowstone has a snowmobile problem—forty years ago the superintendent thought that would be a grand way to increase his winter visitation. Now this—just at entrances and public facilities, they say, but once they open the door they'll be wiring the backcountry because the pressure to do so will be immense. The time to stop that is now."

I think he is absolutely right. Once the door is opened, it is tough to close. I well remember Yellowstone superintendent Jack Anderson proudly opening the park to snowmobiles and receiving an award of merit for doing so from the International Snowmobile Manufacturers Association. Anderson vowed the snowmobiles would be an asset and do no damage. Now, from mid-December to mid-March Yellowstone bans automobiles from most of the park, but it welcomes snowmobiles to travel on many miles of snow-covered roads. In each of four separate studies since 1998, costing a cumulative $10 million, the National Park Service has verified conclusively that greater volumes of traffic required by an emphasis on snowmobiling add dramatically to air and noise pollution and disturbance of Yellowstone's wildlife. The use of snowmobiles has made Yellowstone a chronic, unending legal battleground between the protectors of park values such as tranquility and solitude and the snowmobile lobby, local chambers of commerce, and the politicians they influence.

In the late 1970s Philip Iversen, superintendent of Glacier National Park, denied snowmobile access, despite congressional pressure. He succeeded and that pressure dissolved and disappeared. The worst step is the first step. As I said earlier, when the doors open, they are tough to close. National parks are at their best when those in charge stick to the rules.

45

Going Home

One day, while at Harpers Ferry National Historical Park in West Virginia, I walked to the favorite vista point of Thomas Jefferson overlooking the scenic Potomac and Shenandoah Rivers. There, at the landmark known as Jefferson Rock, I encountered a young boy, about ten or twelve, carving his name into the rock, with the encouragement and direction of his father standing next to him. When I told them this was not the proper thing to do, the father replied, "But a lot of other people have done it, too." That was true, but it did not make it right.

Another time, in the Great Smoky Mountains, I stood behind a wire fence heeding a sign warning visitors to keep out and to leave the wildlife alone. This did not deter a young family of four determined to film a deer with their movie camera. I watched a young fellow, a private citizen, chase that family back to where it belonged, which definitely made them unhappy.

This sort of thing is common in national parks. A favorite ploy of young parents is to try to place a child on the back of a bear for a photograph. My friend Verne Huser, writing from the Tetons, reported seeing flashing road signs that ask drivers to slow down and warn of the potential for collisions with wildlife, but many drivers speed through the park or stop in the middle of the roadway when they see wildlife. Rangers do their best to teach visitors a little common sense, but there are too few rangers to get the job done.

"We need more education," Huser wrote, "perhaps requiring visitors to watch a video before entering the park."

This leads me to cite a statement of Stephen Mather: "The American public possesses an empire of grandeur and beauty which it scarcely knows. It owns the most inspiring playgrounds and the best equipped nature schools in the world and is serenely ignorant of the fact."[1]

And to cite this 2010 letter from Steve Voorhis, director of the Tremont Environmental Center in the Great Smoky Mountains:

> In a park with nine million visitors a year it may not seem significant that we serve less than 6,000 people a year. Too often quantity seems to be the more noticed metric than quality. It is easier to measure quantity so perhaps that is part of the reason.
>
> I often describe that what we do through residential experiences at Tremont, is at the far end of the spectrum of meeting visitor needs, where we influence a handful of people but we influence them very deeply. The students (of all ages) who come through our programs step away from their busy and modern existence and live, learn, play, observe, investigate, explore, study, empathize with, and drink deeply from Great Smoky Mountains National Park. Some more than others, but people leave feeling like they've experienced this place, not just passed through it. They have a relationship with the Great Smokies and it has become more "their park" than before they came. When they come back, they are returning home.
>
> While we teach about the issues and threats facing the Smokies it is more important that we first help people to become connected to this place. It is only when people feel ownership that they will take notice of the threats and issues and hopefully act on them. But even if those people don't act, the power of the parks to, as you said, "transform the human spirit," is needed today more than ever. Opportunities to help them do so should remain an important objective. National Parks have always been about people, their relationship to a place, and the way that those places and *the scenery and natural and historic objects and the wildlife therein* have influenced them deeply. That is what I believe Tremont, North Cascades Institute and our sister organizations and kindred spirits in Yosemite, Cuyahoga, Delaware Water Gap, Olympic and many other parks in varying degrees, are about. I believe what we do will continue to become more and more necessary. People need to be shown how to slow down, to observe, to contemplate, to walk slowly and breathe deeply and marvel at the wonders that are preserved in our parks. We can afford people the chance to truly step away for a time and show them the secrets and details that they might have passed by.

There is much to what he said. The value of the national parks is often defined in terms of numbers of visitors and the amount of money they bring to local communities, or, on another level, the diversity of salamanders, flowering plants, and trees; but all of these are material matters and, for the most part, measurable. Deeper questions worth pursuing are the influence of a natural environment as an educational document, a source of learning with depth and scope about the human environment, the human self, and human dignity. We must learn to embrace a lifestyle that treads lightly on the garden, our earth, and know that doing so will add richness to our lives.

I regard each national park as a sanctuary to transform the human spirit. I feel uplifted and inspired by the places I have visited and by many of the people associated with them.

Yes, the episodes cited above suggest that many Americans have been widely separated from nature and wildlife. But I believe that each person needs and deserves his or her own sacred mountain. I visualize a national park as my sacred mountain even when there is no mountain at all. The sacred mountain speaks as a place of spirit, counseling that transforming society begins with the person. It is a place to recognize the limitations of a fragile earth, to pledge allegiance to a green and peaceful planet, and to believe that if others do likewise and we believe strongly, we will make it happen.

Laws and regulations have their place, but only people make things work. Democracy is what we make of it, a system under which we the people get what we deserve and what we demand. What is most needed, as Horace Albright expressed years ago, is "wider support from more citizens who will take the trouble to inform themselves of new needs and weak spots in our conservation program."[2]

At a 1984 symposium on Hawaiian ecosystems, a report by two scientists, Alan Holt and Barrie Fox, acknowledged as follows:

> If a thousand years from today there are large areas of native landscape in Hawaii, it will be because people cared enough to save them, cared enough to keep natural areas protected even in the face of other potential uses of these lands. The long-term success that we all hope for depends on the people's appreciation of the land. The best prospect for making that future happen is to show today's people the value of our natural heritage and to show them how to care for it.

Americans need to learn the value of our heritage and to show ourselves responsible for it. The best way, if you ask me, is to live lightly on the land, to consume less of earth's goodies, and to conserve more of them. Rather than the present overconsumption and waste, Americans need to restore America the beautiful, from sea to shining sea, and to make peace with the wild species that suffered in the Gulf of Mexico, and elsewhere in the country as well. We as humans need to recognize that other species have their rights, and we have our responsibilities for them.

But this has not happened. It is widely recognized that parks are overcrowded, underfunded, and politicized; that too many automobiles cause congestion and pollution; that off-road vehicles and low-flying airplanes and helicopters pollute the wilderness with noise; that disruptions of park values are commercial in nature and manifest in politics; and that many visitors wander off the established paths, scarring the landscape and leaving litter in their wake.

For a time I thought the way to solve this problem would be to remove national parks from the Department of the Interior and reconstitute the National Park Service as an independent bureau, like the Smithsonian Institution. On reflection, however, that may not be the answer. Parks ought to be a public issue. Perhaps Abraham Lincoln said it best in 1858: "Public sentiment is everything. With public sentiment, nothing can fail. Without it, nothing can succeed. Consequently he who molds sentiment goes deeper than he who enacts statutes or pronounces decisions."[3]

Certainly the issues of preservation versus use, and of ecology versus crowds and commerce, will be subject to continued controversy. But the place to begin is to recognize that conservation responds to social needs. It treats ecology as the economics of nature, in a manner directly related to the economics of humankind. Keeping biotic diversity alive is the surest means of keeping humanity alive. But conservation transcends economics—it illuminates the human condition by refusing to put a price tag on the priceless.

No matter how much experts try to measure "specific benefits," there really is no way to place a dollar value on a park experience. Each national park serves as an outdoor museum of natural and human history, a laboratory, a field classroom, and yet its greatest benefit is in sheer inspiration beyond measure. There is a sense of pride in country, too—that we have all agreed to set aside these sanctuaries to serve the common good. Public parks, like art galleries, theaters, museums, and libraries, enrich society in immeasurable ways; they enlighten and elevate individuals who come to them.

Nature belongs where people live. The more of it—in city, county, and state parks—the better the quality of community life. Open spaces with landscaping, fountains, green sitting areas, and sculpture give cities heart and soul. Public space makes even the poorest citizen wealthier. Parks encourage people to conserve at home and in their neighborhoods, and once they make a difference, community by community, then hopefully in time the scale expands to a larger, global dimension.

I hope I have shown in these pages that a national park is a different kind of park. National parks approach the last representation of primeval life. In a setting free of human intervention, the visitor absorbs the "feel" of nature—of plants, animals, and natural features, and the weathering of the land by winds, rivers, and other geological forces. Intangible values prevail: beauty, timelessness, solitude, silence, harmony, awareness, simplicity, freedom, balance, and order—the essences of humanity at its most humane.

I see natural sanctuaries as human sanctuaries, reminders that we are all connected, as brothers and sisters of common origin, common destiny. I see the whole world, the rich and poor of it, plagued in my lifetime by global warming, acid rain, destruction of tropical forests, loss of wildlife, toxic wastes, poisoned air and water, and the pressures of a growing population.

But I also see that concerns for all, human and nonhuman animals, have come into our moral gaze. I doubt that we shall ever see a perfect world, but we can all take heart from the spirit of our parks, joining together toward that goal. Compassion must be at the root of values; the power of human life is in emotion, in reverence and passion for the earth and its web of life. A feeling, a philosophy, and a love of earth count most. To identify with life on the green planet in all its forms, as we do in our national parks, is to celebrate human hope and human potential.

Crowds, congestion, noise, intrusions of human-made structures, and pollution of air and water all interfere with the appreciation, understanding, and enjoyment of the natural scene. To be fully enjoyed, the parks must first be fully preserved.

A slower pace expands the dimensions of time. It enables us to recognize the limitations of a fragile earth, to pledge allegiance to a green and peaceful planet, and to believe that if others do likewise and we believe strongly, we will make it happen.

Notes

Preface and Acknowledgments

1. John Muir, *A Thousand-Mile Walk to the Gulf* (Boston: Houghton Mifflin Company, 1916).
2. John Muir, *Our National Parks*, Foreword by Richard F. Fleck (1901; reprint, Madison: University of Wisconsin Press, 1981).

Chapter 1. Feet on the Ground, Eyes to the Sky

1. John Muir, *Steep Trails* (Berkeley: University of California Press, 1918).

Chapter 4. A Superintendent's View of Park Priorities

1. Michael Frome, *Whose Woods These Are: The Story of the National Forests* (New York: Doubleday, 1962), 288.

Chapter 8. John Muir Comes into My Life

1. See Tom Wallace, "Victory for Cumberland Falls," *American Civic Annual*, Vol. 2 (Washington, DC: American Planning and Civic Association, 1930), 141–245.

Chapter 9. No Glaciers or Geysers in the Everglades, but . . .

1. Dan Beard, "Preliminary Thoughts on the Master Plan" (unpublished report, 1938), 100.

Chapter 11. I Become an Author

1. Henry David Thoreau, *The Maine Woods* (Princeton: Princeton University Press, 1983).
2. William Cullen Bryant, "Thanatopsis," *Yale Book of American Verse*, ed. Thomas R. Lounsbury (New Haven: Yale University Press, 1912), 36–38.
3. William Cullen Bryant, "A New Park," *New York Post*, July 3, 1844.
4. Pierre Charles L'Enfant, report to George Washington.

CHAPTER 12. HORACE ALBRIGHT

1. Mather was principled and a true believer, asserting with pride that national parks were "America's most inspiring playgrounds and best equipped nature schools." *National Parks Portfolio* (New York: Charles Scribner's Sons, 1916), 1.

   John Ise describes Mather's administration of the fledgling National Park Service, from 1917 to 1928, in *Our National Park Policy: A Critical History* (Baltimore: Johns Hopkins University Press, 1967).

   But he was also a master marketer, a pitchman with class who proceeded to promote the parks like a product, encouraging Americans to visit and explore the far-flung natural wonders. He saw promotion as the expeditious way to build a public constituency to influence Congress to support the bureau, enhance its prestige, and add new areas to its jurisdiction.

   Mather's family was from New England, but he was born and raised in California. Following graduation from the University of California, Berkeley, Mather worked for five years as a reporter for the *New York Sun*. Then he went to Chicago to do advertising and sales promotion in the borax business, which looked so attractive that he mastered finance and acquired a healthy piece of the borax market. The story has been told of how, after visiting the national parks in 1914, he complained to Secretary of the Interior Franklin K. Lane (whom he had known as a fellow student at the University of California) about how poorly they were run. Lane was an experienced public administrator and political progressive who hoped to streamline and modernize the Interior Department, and so he challenged Mather to come to Washington to run the national parks for himself.

   In 1917, Mather, a Californian, became one of the founders of the Save-the-Redwoods League, working for the preservation of California's majestic coast redwood groves. (See John B. Dewitt, *California Redwood Parks and Preserves: A Guide to the Redwood Parks and a Brief History of the Efforts to Save the Redwoods* [San Francisco: Save-the-Redwoods League, 1982].) Others included Lane, as the league's first president; Representative William Kent; John C. Merriam, professor of paleontology at the University of California (later president of the Carnegie Institution in Washington, D.C.); Henry Fairfield Osborn, president of the American Museum of Natural History; and Madison Grant, president of the New York Zoological Society.

   The league was a private organization oriented to private responsibility and local administration. Its leaders included scientists and scholars, gentlemen of esteem, some of them close friends of Herbert Hoover, who decried coercive regulatory bureaucracy. They believed in the lofty obligation to do good for their golden state through moral persuasion and cooperation, equally with lumber companies, state officials, conservationists, and philanthropists, which explains why the league long rejected proposals for a redwood national park. Thus in 1921, the Kent-Mather Grove was acquired, with funds donated by two of the league's eminent pioneers. So things were done in those days, with little of the acrimony and confrontation that characterized efforts during the 1960s and 1970s, when a few great redwood groves were saved while others were destroyed by the logger's chainsaw.

Mather kept his borax interests and continued to draw a salary and share of the profits, which he dispensed generously on behalf of the national parks. He wined and dined those who counted, entertained influential editors on field trips, and hired an old crony from his newspaper days, Robert Sterling Yard, as publicity chief of the Park Service, paying Yard's salary from his own funds (until 1918 when Congress forbade the payment for government work by private funds).

He had great moments as a leader, with courage to dare and defy politics and politicians. Early on, for example, he fought a plan of Idaho irrigation advocates to draw water from the southwest corner of Yellowstone and from Yellowstone Lake. It was one of many incursions he resisted. In a 1920 report titled "A Crisis in National Conservation," Mather wrote: "Once a small dam is authorized for irrigation or other purposes, other dams will follow. Once a small lake is raised and a small amount of timber is destroyed . . . once start the national parks toward national forest-status, and it will be logically impossible to stop short. One misstep is fatal." Robert Shankland, *Steve Mather of the National Parks* (New York: Alfred A. Knopf, 1954), 215–16.

2. The intent is evident in this passage: "Congress declares that the national park system, which began with the establishment of Yellowstone National Park in 1872, has since grown to include superlative national, historic, and recreation areas in every major region of the United States, its territories, and island possessions; that these areas, though distinct in character, are united through the inter-related purposes and resources into one national park system as cumulative expressions of a single national heritage; that, individually and collectively, these areas derive increased national dignity and recognition of their superb environmental quality through their inclusion jointly with each other in one national park system preserved and managed for the benefit and inspiration of all the people of the United States; and that is the purpose of this Act to include all such areas in this System and to clarify the authorities applicable to the system." National Parks Act, August 25, 1916.

3. The protection of scenic beauty in national parks, however, represents only one phase of Albright's career. His touch was felt in regional and city planning through the pioneering work of the American Planning and Civic Association, of which he was president for a score of years, and also in the movement to restore historic places, through his active concern over George Washington's birthplace, Colonial Williamsburg, and Jamestown Island, scenes where the story of a nation began.

Albright served on the board of directors at Colonial Williamsburg for twenty-four years, almost from the very start. During a visit to

Williamsburg in the 1950s, I learned from A. Edwin Kendrew, the senior vice president, whose career then spanned its entire history, of the part played by Albright. Kendrew recalled:

> Getting Williamsburg going was daring in its way, like plowing new ground. There was nothing like it in America, nor dreamed of. But Mr. Rockefeller was not interested in restoring a single building in inappropriate surroundings—his objective was to revive an entire segment of the past or none at all. He was reticent, cautious and kept largely to himself. But Horace Albright was able to penetrate with sound advice in particularly trying areas—of principle, policy and ideals, and of spending money in the right places.
>
> Looking back, it is difficult to define Horace Albright's precise role at Williamsburg. He was helpful in zoning and city planning in our painstaking, slow endeavor—the battle to turn back the pages of history with authenticity. He supported the architects and other professionals in matters which some considered visionary, if not unnecessary, in the Twentieth Century context. One was the instance of setting the new hotel well back from the street. Or, he would tell Mr. Rockefeller that detailed research into the shape and substance of original buildings might be costly, but was essential. He championed purchase of land not only for restoration but for protection of the area from encroachment. "It will never be any cheaper, you know," he would suggest to Mr. Rockefeller. When it was proposed to intrude on history with modern street lights, he insisted that one concession would only lead to another. "Instead of giving the visitor convenience," he said on one occasion, "we should give him a street map."
>
> He could make his points stick, at Williamsburg and elsewhere, because he had a love for people at all stations, for a carpenter or gardener no less than for Laurance Rockefeller, whom he influenced deeply to take up his father's work in conservation. In time, when Congress passed the Historic Sites Act and the Park Service undertook its own restoration work, Washington people would say, "Let's go down and see how the experts do it at Williamsburg." That fine rapport was his work.

CHAPTER 13. DRURY IN DEFENSE OF DINOSAUR

1. *National Parks Magazine* 97 (April–June 1949): 28. Cited in John Ise, *Our National Park Policy: A Critical History* (Baltimore: Johns Hopkins University Press, 1967), 7.
2. Ise, *Our National Park Policy*, 475.
3. Washington's loss was California's gain. Governor Earl Warren, Drury's old classmate, forthwith appointed Drury director of California's Division of Beaches and Parks, which he had been instrumental in establishing years before. The system embraces natural areas from the Anza-Borrego desert to

redwood rain forests, sorely needed beach parks along the Southern California coast, and historic parks recounting virtually every phase in human history, including homes of figures like Jack London and Will Rogers, and the San Simeon castle of William Randolph Hearst. In 1959 Drury reached the state's mandatory retirement age of seventy, then served again as executive secretary of the Save-the-Redwoods League, then later as president and chairman.

CHAPTER 14. HETCH HETCHY AGAIN, NOW IN UTAH

1. John B. Oakes, "Conservation" column, *New York Times*, March 4, 1951.
2. Richard C. Bradley summed it up in the Sierra Club Bulletin of December 1964: "This began as an internal dispute within the Interior Department, but eventually spread from coast to coast and brought to government officials and congressmen more mail than they had ever received on any single issue. The battleground was a peaceful little valley in western Colorado where the Yampa and Green rivers come together beneath Steamboat Rock—as little known to the public in 1950 as Guadalcanal had been in 1940. Within a few years Echo Park would appear on the editorial page, if not the front page of major newspapers all over the country." Richard C. Bradley, "Grand Canyon of the Controversial Colorado," Sierra Club Bulletin, December 1964, 74.

CHAPTER 15. WHOSE WOODS THESE ARE

1. Michael Frome, *Whose Woods These Are: The Story of the National Forests* (Garden City, NY: Doubleday and Company, 1962), back cover.
2. To a large degree, this legislation reflected the active influence of Pinchot. Born to wealth in Pennsylvania the last year of the Civil War, he graduated from Yale in 1889 and set sail for France to study forestry (since there was no place in America offering any such course of study). On his return, Pinchot was engaged, on the recommendation of Frederick Law Olmsted, to develop a forest plan for the George Washington Vanderbilt estate at Asheville, North Carolina. Olmsted was the renowned landscape architect of his day, who had helped design Central Park in New York and then, for a time, pursued further work in California on the way to his career in designing city parks across the country. Olmsted saw preservation of natural scenes as beneficial to the human spirit, as well as for recreation. Pinchot, for his part, was bent on proving that forestry deserved a place in America, based on his belief that productive lands and an abundance of resources determine the quality of living of any nation. Forestry, to him, was an issue of public welfare.
3. John Muir, *The Mountains of California* (New York: The Century Company, 1907), 244, 256.
4. Robert Engberg, ed., *John Muir: Summering in the Sierra* (Madison: University of Wisconsin Press, 1984), xiii.

5. John Muir, *John of the Mountains: The Unpublished Journals of John Muir* (Madison: University of Wisconsin Press, 1979), 351.
6. Gifford Pinchot, *Breaking New Ground* (Washington, DC: Island Press, 1998), 103.
7. Roosevelt made this evident when he addressed a joint session of Congress in December of that year. In defining a new era of national unity and growth, in words that might easily have been drafted by Pinchot, he declared: "The forest and water problems are perhaps the most vital internal problems of the United States. The fundamental idea of forestry is in the perpetuation of the forests by use. Forest protection is not an end in itself; it is a means to increase and sustain the resources of the country and the industries which depend on them. The preservation of the forests is an imperative necessity."
8. See Robert Shankland, *Steve Mather of the National Parks* (New York: Alfred A. Knopf, 1954), 49.
9. Message from President Taft to Congress, February 2, 1911; cited in Shankland, *Steve Mather of the National Parks*, 52–53.
10. Frome, *Whose Woods These Are*, 181.
11. Aldo Leopold, *A Sand County Almanac, and Sketches Here and There* (New York: Oxford University Press, 1989), xxvi.

CHAPTER 16. ON BECOMING A COLUMNIST

1. Morris K. Udall, *Too Funny to Be President* (New York: Henry Holt and Company, 1988), 59–60.
2. Brower charged at that symposium that the Sierra Club book on the Grand Canyon was being suppressed and was not available for sale in the national park. Howard Stricklin was the park superintendent at that time, and his feathers were ruffled. "Why, of course we have the book for sale. It's right here," he said only an hour after Brower had made his remark. And there it was, hidden under the counter. Secretary Udall was one of the conservation heroes of the period, but a principal advocate of the proposed dams on the Colorado River and of environmentally destructive power development in the Southwest. As a consequence, national park people were silent and silenced.

CHAPTER 17. SPEAKING AT YALE

1. Gifford Pinchot, *Breaking New Ground* (Washington, DC: Island Press, 1998), 509–10.
2. Gifford Pinchot, *The Fight for Conservation* (New York: Doubleday, Page & Company, 1910), 115–16.
3. Ibid., 20, 123.

CHAPTER 18. THE TIMID, THE HESITANT, THE COMPROMISERS HAVE FAILED

1. I will say, however, that more than any other secretary of the Interior before or after his time, Harold L. Ickes, in the years of Franklin D. Roosevelt's presidency, enunciated a clear policy of preservation and committed himself to make it work. At the superintendents' conference in February 1936 he expressed this philosophy on the administration of national parks:

> I do not want any Coney Island. I want as much wilderness, as much nature preserved and maintained as possible. If I could have my way I would have much fewer roads in most of the parks.
>
> I recognize that a great many people, an increasing number every year, take their nature from the automobile. I am more or less in that class on account of age and obesity. But I think the parks ought to be for people who love to camp and love to hike and who like to ride horseback and wander about and have a community of interest, a renewed communion with nature.

CHAPTER 19. "PARKS ARE FOR PEOPLE" MAKES THE GREAT SOCIETY LOOK GOOD

1. In his welcoming letter to the First World Conference on National Parks in Seattle in 1962, Kennedy wrote: "We must have places where we can find release from the tensions of an increasingly industrialized civilization, where we can have personal contact with the natural environment which sustains us. . . . The permanent preservation of the outstanding scenic and scientific assets of every country, and of the magnificent and varied wildlife which can be so easily endangered by human activity, is imperative." John F. Kennedy, welcoming letter to First World Conference on National Parks, Seattle, WA, June 30–July 7, 1962. Edited by Alexander B. Adams. National Park Service, United States Department of the Interior (Washington, DC: U.S. Government Printing Office).
2. In 2010 Joe Browder, long active in defense of Everglades National Park, wrote to me as follows:

> Udall was on the wrong side of Everglades. After he left Interior and started his consulting group, his first big client was the Dade County Port Authority, trying to help them sell the Everglades Jetport. Udall hired as his Everglades consultant a dear friend of mine, Frank Craighead (father of the Craighead brothers). When he ordered Craig not to communicate with me, Craig threatened to resign, Udall kept him, Craig wrote a report saying that Everglades National Park and Big Cypress itself would be devastated by the jetport. So Udall tried to

> finesse things. He began whispering about a jetport combined with federal/state planning that would let some of Big Cypress be responsibly developed, while preserving a few of the larger cypress stands. Eventually he gave up on the jetport, but still supported the planning/zoning, which became the developers' (and Hartzog's) preference. We whipped their tails.

3. Comptroller General of the United States, "Effectiveness of the Financial Disclosure System for Employees of the U.S. Geological Survey," March 3, 1975.
4. See William K. Wyant, *Westward in Eden: The Public Lands and the Conservation Movement* (Berkeley: University of California Press, 1982). Wyant discusses the Teapot Dome scandal in chapter 4, "The Fall of Albert B. Fall."
5. See Barry Mackintosh, "Harold L. Ickes and the National Park Service," *Journal of Forest History* (April 1985): 78–84; Donald C. Swain, *Wilderness Defender: Horace M. Albright and Conservation* (Chicago: University of Chicago Press, 1970); Harold L. Ickes, *The Autobiography of a Curmudgeon* (New York: Reynal and Hitchcock, 1943).
6. Address of Harold L. Ickes, Superintendents' Conference, Washington, DC, February 1936. Copy provided by Louise Murie from the files of Adolph Murie.
7. The Park Service released its report, "The Redwoods: A National Opportunity for Conservation and Alternatives for Action," in September 1964. "The proposal suggested three different plans for a park on Redwood Creek varying in size from 30,000 to 50,000 acres." Susan R. Schrepfer, *The Fight to Save the Redwoods: A History of Environmental Reform, 1917–1978* (Madison: University of Wisconsin Press, 1983), 121.
8. See Schrepfer, *The Fight to Save the Redwoods*, chapter 8, "The Redwood National Park: 1965–1968."
9. Udall wanted to spark the conservation crusade and be recognized for it, but he was pragmatic in the most political sense of the word. Thus, on September 26, 1968, he went to dedicate the visitor center at Glen Canyon, the dam that conservationists even now despise above all others. He would be embarrassed by the words he spoke that day, and later would say that they were a mistake on his part, but his statement belongs in the record:

> Before the Glen Canyon Dam was built the river ran through here red with sediment, and the back-country was a no-man's land except for a handful of shepherds and those fortunate few visitors who were able to make their hazardous way through nearly impassable rugged terrain to Rainbow Bridge and other marvels of the side canyons.
>
> Today, thanks to the dam, the Colorado River now runs clear, and boaters on jewel-like Lake Powell may float up hidden canyons and feast their eyes on some of the most fantastic and gorgeous scenery in the world.

> We are proud that Glen Canyon Dam was awarded the American Society of Civil Engineers award for the outstanding civil engineering achievement of 1963. And while the Glen Canyon unit furnishes us superb scenery, glorious boating, and excellent fishing, it is also generating hydroelectric power that will pay for construction of the facilities for other basin developments to come.
>
> The $832 million Central Arizona Project will pump water from Lake Havasu on the Colorado River and transport it by aqueduct some three hundred miles to the burgeoning Phoenix-Tucson area, one of the fastest growing sections in the Nation. We are all aware of the fabulous expansion of the sun-drenched area during the past decade and of the prospects for even greater growth in the future Reclamation has made the past progress possible, and this Reclamation project will help it to continue.

U.S. Department of the Interior, press release, "Excerpts of Remarks by Secretary of the Interior Stewart L. Udall at Glen Canyon Visitor Center Dedication, September 26, 1968."

10. Donald W. Carson and James W. Johnson, *Mo: The Life and Times of Morris K. Udall* (Tucson: University of Arizona Press, 2004), 134.
11. Morris K. Udall, "Flooding the Grand Canyon: A Phony Issue," *Congressional Record*, June 9, 1966, Morris K. Udall Papers, Special Collections, University of Arizona Library, Tucson.
12. Robert Ashton Jr., Richard Endress, Diane Traylor, and Bruce Panowski (Interpretive Ranger Division, Mesa Verde National Park), press release, September 23, 1970, Boulder, Colorado.

CHAPTER 20. BUILDING AN EMPIRE THROUGH POLITICAL PATRONAGE

1. Coalition of National Park Service Retirees, "NPS Retirees: Yellowstone Should Comply With Court Order by Providing More Snowcoach Access and Phasing Out Snowmobile Use in Two Years," September 15, 2008.
2. Ibid.
3. Biographical data: George B. Hartzog Jr., July 1971. Sent to the author by U.S. Department of the Interior, Office of Information, on March 23, 1972.
4. John McPhee profiled Hartzog in a *New Yorker* article titled "Ranger" (published on September 11, 1971), though Hartzog was never a ranger. McPhee wrote of Hartzog's central staff, his high command, but without providing much information about their training, personal outlook, or goals. The article said that of the seven Park Service directors, Hartzog was the second to come up from ranger, without identifying the first (and neither do the records).

   McPhee told a revealing story of Hartzog's approach to green space. Office buildings were coming down on the National Mall in Washington, which the Park Service administered. "The last thing we need in downtown

Washington is more grass," Hartzog complained. "We've got grass coming out of our ears in this city, and in summer we let it turn brown. We're up to our noses in horticulturists who don't know enough not to water grass when it gets hot. We need more vistas like a Buick needs a fifth hole." It was consistent with his attitude toward preservation of nature. It was not what he favored most.

5. As an example of how well he stood behind his employees, Hartzog cited the particular case of Roger Allin, superintendent of Everglades National Park, who took a lot of political heat for efforts to ensure the flow of water to the park: "Roger did a superb job," wrote Hartzog. Nevertheless, he continued, Senator Spessard Holland, who had helped to establish the park, demanded, "Move this Allin guy out of the Everglades." "I allowed as how important he [Holland] was to the park, but declined to move Allin, saying to the senator that Allin was doing exactly what I had asked him to do." George B. Hartzog Jr., *Battling for the National Parks* (Mt. Kisco, NY: Moyer Bell Limited, 1988), 228–29.

   A 1985 taped interview with Roger Allin by the author recorded the following:

   > "Was it George who decided it was time for you to leave the Everglades?"
   >
   > "Oh yes, there's no question about that. He'd been at me for about six months and finally told me to come up to Washington; he wanted to talk to me. When I came he said, 'I'm just telling you eyeball-to-eyeball that you're coming up here.' I think I had done his job, I had done a good job, was recognized as a competent manager, and wanted to stay—very, very badly. But he told me that was where I was going to pick up my paycheck, and so I went."

   Allin went further:

   > "He had a magnificent command of the language, was enormously talented and persuasive, but didn't have much concern really for the people working under him. I think that his grandmother, if she wasn't sold down the river, was fortunate. He was an absolute master administrator and I had the highest admiration for him, but I didn't think much of him as a person."

   Interview with Roger Allin, Whidbey Island, Washington, 1985.

6. In 1975, the Justice Department investigated Hartzog and a former Interior Department lawyer, Bernard R. Meyer, for possible conflict of interest. The previous fall they had become lawyers for Landmark, operator of tourmobile buses in Washington, less than two years after leaving government. They had met with Interior Department officials in efforts to negotiate a

new twenty-year contract for the firm. They resigned as Landmark's lawyers and as lawyers for MCA, the parent company. "I don't see what else I could have done besides practice law," Hartzog was quoted as saying (*Washington Post*, August 3, 1975), and the legal area he knew best, as he said, was the federal parks.

He was involved in real estate as well. He was a twenty percent partner (with Gerald Halpin, a prominent developer in northern Virginia with Democratic party connections) of a multimillion-dollar real estate deal at Harpers Ferry, West Virginia. The plan called for construction of a two-hundred-room hotel and one hundred and twenty townhouses on forty acres of cliff-side property being considered for acquisition as an addition to Harpers Ferry National Historical Park. The plan proposed exactly what the Park Service was trying to prevent: commercial development along the Potomac River. The feasibility of construction on the steep hillside covered with rock outcroppings was dubious, but it might have resulted in a substantial increase in the property value.

The *Hagerstown Morning Herald* of August 8, 1979, reported that Hartzog dropped out of the deal after receiving a pointed letter from Park Service director William Whalen: "We would strongly recommend that you withdraw the application so that rumors and gossip will be stopped. . . . [Y]ou would be ill-advised to pursue this action in the context of our expressed desire to acquire that property." Hartzog had claimed he knew nothing of the Park Service interest in the land when he purchased it, but Whalen wrote: "I would like to point out that the authorized boundary of the park was authorized in 1974 and was public knowledge throughout the community prior to your involvement. I would also like to point out that the legislation that expanded the boundary in 1974 was based on studies that were generated as early as 1971 and completed early in 1973."

In 1971 Hartzog granted a right-of-way across Gettysburg Battlefield to Thomas R. Ottenstein, a businessman in suburban Washington, in return for three acres conveyed to the National Park Service, at the same time giving clearance for the construction of a three-hundred-foot tower that was subsequently widely denounced as an "environmental insult." The state of Pennsylvania entered legal action to stop what Governor Milton J. Shapp called "the second battle of Gettysburg." (U.S. Department of the Interior, National Park Service, press release, July 11, 1971; *Los Angeles Times*, April 4, 1971; *New York Times*, May 16, 1971; Commonwealth of Pennsylvania, Office of the Governor, Harrisburg, press release, July 20, 1971; *Harrisburg Patriot*, July 26, 1971.) Hartzog offered an explanation about being out of town, not knowing what he was signing. (A National Park Service news release dated July 11, 1971, quoted Hartzog in the first paragraph as announcing the agreement, then in the second paragraph referred to Interior Department special assistant J. C. Herbert Bryant Jr., a political appointee, as the negotiator with Ottenstein, the promoter of the tower.

See Bill Richards, "Tower Power," *Washington Post*, October 28, 1973. Hartzog is quoted as follows: "I O.K.d the agreement by phone from the Grand Canyon. When I got home and saw that access I tried to cancel it but it was too late.")

In letters, signed and unsigned, individuals in the National Park Service expressed bitterness and disillusionment. For example: "Those who show concern for ideas that are being left behind are made to feel like soldiers in a guerilla army who must meet in secret and discuss their convictions only with a trusted few—with irreparable damage to career ambition the inevitable result of exposure."

CHAPTER 22. "WILDERNESS IS MY LIFESTYLE"

1. In 1973 National Geographic produced a high-quality oversize book, *Wilderness U.S.A.*, for which the editors sent me around to report on wilderness in the East. The lead essay in the front of the book, titled "A Longing for Wilderness," was written by Sigurd Olson, who was both a hero and friend of mine. In this essay he wrote:

> The world we face now is a strange one, the great silences replaced with clamor, the hearts of our cities garish with blinking neon and foul with the stench of pollution. We look at the slums, at the never-ending traffic, the shrinking space and growing ugliness, and are appalled. Is this, we ask, what our forebears struggled for? Is this the great American dream?

For my part, National Geographic gave me a generous expense account and allowed me to visit areas of my own choice in diverse locations of the East. These included wild areas under different jurisdictions: Baxter State Park and Allagash Waterway, Maine; Great Gulf, White Mountain National Forest, New Hampshire; Adirondack Forest Park, New York; Great Swamp National Wildlife Refuge, New Jersey; Cranberry Backcountry, Monongahela National Forest, West Virginia; Sipsey Wilderness, Bankhead National Forest, Alabama; Okefenokee National Wildlife Refuge, Georgia; and Everglades National Park, Florida.

A ranger and I explored a portion of the Everglades' watery wilderness by canoe, the only practical way to take an intimate look. We paddled the Hells Bay Trail, one of four trails linking a chain of shelters. The watery trail at times was no wider than the canoe. Mangroves fifty to seventy-five feet high stood on stilt roots arching out and down like legs of giant crabs. I felt that we were swallowed by mangrove forest while tunneling through the tangled thickets. Waters darkened by tannic acid from the mangroves mirrored the jungle in striking clarity. A heron fishing at water's edge broke the silence with a squawk. At thicket openings we caught site of egrets and ibises on the mud flats, the dark back of a limpkin barely visible among mangrove roots.

Another day in the Everglades, park superintendent Roger Allin took me on a ride over the one-hundred-mile-long "wilderness waterway." He was proud of it, especially the title he had given it. But I was not so impressed, and wrote in *Wilderness U.S.A.*:

> It takes seven days to cover the park's Wilderness Waterway by canoe. But it took us only one day in our powered patrol boat. I wondered later about the place of power: Is it possible, really, to justify a man's presence in wilderness unless he gets into it under his own steam? Traveling by machine makes a man part of the machinery, rather than a part of nature; he detracts more than he contributes.

Allin and I became good friends and, in time, he may have thought otherwise about his motorized wilderness trail, but it was typical of the way agency people looked at wilderness and their responsibility to it.

2. George B. Hartzog Jr., "The Wilderness Act and the National Parks and Monuments," in *Wilderness and the Quality of Life*, Proceedings of the Sierra Club Biennial Wilderness Conference, San Francisco, April 7–9, 1967, ed. Maxine E. McCloskey and James P. Gilligan (San Francisco: Sierra Club, 1969), 17.

Chapter 27. Concession Power

1. James Ridenour, *The National Parks Compromised: Pork Barrel Politics and America's Treasures* (Merriville, IN: ICS Books, 1994), 133.
2. Ibid., 136.
3. Joseph L. Sax, *Mountains Without Handrails: Reflections on the National Parks* (Ann Arbor: University of Michigan Press, 1980), 12.
4. Horace Kephart, *The Book of Camping and Woodcraft: A Guidebook for Those who Travel in the Wilderness* (New York: The Outing Publishing Company, 1906), 5–6. But concessionaires and today's park administrators want to make things more "comfortable," as part of a cushier, and costlier, experience. In this regard, a friend sent me an editorial published in the *Winston-Salem (North Carolina) Journal* on July 6, 2008, written by editorial page editor Linda Brinson. It was titled "KEEP Roughing It, Now more than ever, people need to get up close and personal with nature," and declared:

> The Blue Ridge Parkway, as many people in this part of North Carolina well know, is a lot more than just a road. It's a "linear" park, part of the National Park System. The road meanders through some of the most beautiful territory on earth, around and over lofty mountains, past forests and fields, beside streams and old farms, and through lush valleys. There are lots of overlooks, trails and places and things to investigate. The idea is for people who come to the parkway to take

their time—the speed limit is 45 mph, and many people drive more slowly—and enjoy the natural beauty.

Almost 20 million people do that every year, more than visit any other unit of the National Park System. It remains to be seen whether gas prices will cut into that number this year, but certainly it's not as if the Park Service needs to fancy up the facilities to attract more visitors.

More important, making changes that would put new barriers between visitors and nature would be a big mistake. We have far too many already. . . .

When you get away from if not "it all" at least a lot of the rush and artificiality of modern life, you gain perspective about humans' place in the world. Much has been written recently about today's children and their lack of connection to nature. . . .

Children today are used to more gadgets and distractions and conveniences than ours were 25 years ago. That's why they need the natural world, without the cushy amenities, even more. Let's leave the Blue Ridge Parkway as it was. Odors and all.

CHAPTER 31. MUIR FOUND THE ICY WILDERNESS "UNSPEAKABLY PURE AND SUBLIME"

1. John Muir, "The Discovery of Glacier Bay," *Century Magazine* (May 1895): 238, 245.
2. John Muir, *Travels in Alaska* (MobileReference, 2010).
3. Ibid.
4. Ibid.
5. Muir, "The Discovery of Glacier Bay," 245–47.

CHAPTER 32. TOURIST BOOMERS LIKE ACTION AND A GOOD SHOW

1. Michael Frome, Letter to the editor, *Courier* [National Park Service house organ], October 7, 1987.
2. "National Parks to Benefit from Stimulus Funds," *RV Business*, May 6, 2009.
3. Joseph Wood Krutch, "Invasion of Baja California," *New York Times*, February 22, 1959.
4. Ibid.
5. Joseph Wood Krutch, *Defenders*, vol. 59 (1984): 37.
6. Chambers of commerce, tourist bureaus, and some local newspaper editors find this hard to believe. Here, for example, is an op-ed piece from the *Traverse City* (Michigan) *Record-Eagle* of Sunday, October 19, 2008:

Sleeping Bear Dunes are so unique, or so the story goes, that astronauts can identify their sandy outline from space.

Here on the ground, however, the Sleeping Bear Dunes National Lakeshore that stretches across Leelanau and Benzie counties is getting to be downright invisible. The lakeshore's visitor numbers plunged this summer, reflecting an estimated 30 percent drop-off in July, and resulting in the park's lowest visitor numbers in 20 years. That's on top of some shaky numbers from 2007, when data from Michigan State University showed the park's visitor numbers fell 6.5 percent from the 2006 totals.

That just doesn't add up when much of the region's tourism industry is holding its own in a tough Michigan and national economy.

Those who've been around long enough remember that locals were sold on the idea of a national park in northwest Michigan because it would be an economic engine for businesses in and around the lakeshore. It also would be a world-class attraction right in our backyard that would be around for our children's children to enjoy.

But on both counts, the feds haven't lived up to their end of the deal. Instead of a primary attraction, the lakeshore evolved into almost an afterthought in the region's tourism scene—something to do when you've done everything else. A national park established in the 1960s is still operated like it's the '60s, with little if any local and regional promotion of the lakeshore, or northwest Michigan in general.

Does it make any sense that Michigan spends $30 million a year to promote state travel while the federal government comparatively spends pennies to promote the lakeshore? The "if you build it, they will come" mind-set of the park's early days is out the window in the 21st century business world, where there's plenty of hot competition for tourism dollars.

The Park Service also violated the public's trust when lawmakers instituted park entrance fees several years ago. Locals and visitors were forced to pay to visit areas they enjoyed free for most of their lives. Paying for amenities like camping or boating out to the Manitou Islands is one thing, but there's no reason the public should have to shell out extra money to climb the sand dunes or motor through the breathtaking Pierce Stocking Scenic Drive. With the federal government paying billions on far-away wars and "rescue" plans for Wall Street, it's silly to charge people to visit property that's publicly owned.

Some of the park's most popular and growing attractions are the private businesses that operate within the lakeshore. Instead of promoting those features, the Park Service for years hassled some of those business owners. The park's current management plan seems more geared toward closing off additional lakeshore areas from the public, instead of convincing more people to visit.

The National Park Service could make significant movement toward reversing its isolationist trends by ending park visitor fees next season, and working more with the local tourism industry to promote itself in Michigan and beyond. It's time for the Sleeping Bear Dunes National Lakeshore to get out of the 1960s, and start making good on some of those long-lost promises its founders made to the community.

CHAPTER 33. THE SCIENTIST WHO SPEAKS FROM CONSCIENCE PAYS A PRICE

1. See Adolph Murie, *The Wolves of Mount McKinley* (Washington, DC: U.S. Government Printing Office, 1971; reprint of National Parks of the United States Fauna Series, no. 5, 1944).
2. The Murie brothers were born in Minnesota, where they camped, fished, swam, and canoed as a prelude to academic study. In 1920 at the age of thirty-one, Olaus made the first of many trips to Alaska for the U.S. Biological Survey (later renamed the Fish and Wildlife Service); from then on he traveled hundreds of miles by dogsled each winter, and by boat or on foot each summer, conducting a biological survey of the Alaska wilderness, with emphasis on the life history of caribou.

   In 1927 Olaus came to Jackson, Wyoming, with his wife, Margaret, who had spent her childhood in Fairbanks and shared his love of wild country. He began his study of elk, which later resulted in the definitive work *The Elk of North America*, and, over the years, led pioneering scientific expeditions to the Aleutian Islands, the Brooks Range (influencing establishment of the Arctic National Wildlife Refuge and, later, Gates of the Arctic National Park) and New Zealand. He was a self-taught artist who tended to details in field sketches—head, ears, mouth, nostrils, legs, the true colors—and left behind a legacy in wildlife art. I have sketches of his work on my wall that enrich our setting. In 1946 he resigned from the government to become director, and then president, of the Wilderness Society. His wife, too, was a gifted writer, the author of *Two in the Far North*, which recounted their adventures together.

   Olaus took a few words that he and his wife found on an old tombstone in Cumberland, England, and reproduced them on a plaque that hung on the mantle of their log home at Jackson Hole: "The wonder of the world, the beauty and the power, the shape of things, their colours, lights and shades—these I saw. Look ye also while life lasts." Those words are a guide to the living; for the Muries they reinforced a perception of nature stronger and richer than scientific analysis, yet invaluable to it. Olaus died in 1963; Margaret, or "Mardy," carried on until 2003 (when she was 101), with considerable influence derived from heartfelt human caring.
3. Foreword to Adolph Murie, *The Mammals of Mount McKinley* (Mount McKinley Natural History Association, 1962), 1.
4. Letter from Mount McKinley Superintendent Duane Jacobs to Adolph Murie, November 15, 1956.
5. Brothers Frank and John Craighead studied Yellowstone's grizzlies for more than a decade, tagging and tracking them with radio transmitters, attempting to quantify the age and sex ratios of the grizzly population and compiling data on social organization, feeding habits, and other aspects of grizzly life history. From 1959 to 1967 the Craigheads collaborated with the Park Service on a friendly basis, enjoying valuable information from park personnel while helping the park manage troublesome bears. The

relationship changed, however, in 1968, when a new Yellowstone Park administration instituted changes in grizzly bear management policy that were "in response to external pressures and in complete disregard of scientific information we had made available," according to Frank Craighead, writing in *Track of the Grizzly*, 1st softcover ed. (San Francisco: Sierra Club Books, 1982), 11.

Thus the research team was in the position of "opposing the official line on management"; over the next few years "the climate for independent scientific research in Yellowstone steadily worsened, and our work was in various ways impeded, misrepresented, and publicly disparaged by park officials because its results did not conform to the changed position of management," Frank Craighead continued. "More important, the new policies were very nearly disastrous to the grizzly community." These events and the ensuing controversy "resulted in the untimely termination of our field work in Yellowstone in 1971."

6. Charlie Ott, interview with the author, 1986.
7. George M. Wright, Ben H. Thompson, and Joseph S. Dixon [U.S. Department of the Interior, National Park Service], *Fauna of the National Parks of the United States: A Preliminary Survey of Faunal Relations in National Parks* (Washington, DC: Government Printing Office, 1933). See Alfred Runte, *National Parks: The American Experience* (Lincoln: University of Nebraska Press, 1987), 139–40.
8. Adolph Murie, *The Grizzlies of Mount McKinley* (Seattle: University of Washington Press; originally published as National Park Service Scientific Monograph Series, no. 14, 1981), 239.
9. Adolph Murie's problems with superiors derived in large measure from publishing independent and critical views in scientific and popular periodicals. *National Parks Magazine* featured his articles deploring expansion of the road in Denali (then Mount McKinley) and pesticide spraying in Grand Teton. His classic, *A Naturalist in Alaska*, was reissued by the University of Arizona Press in 1989.
10. Adolph Murie, "Pesticide Program in Grand Teton National Park," *National Parks Magazine* (June 1966): 17–19. "All this wildness has beauty and harmony for those who have a feeling for wildness. We have set aside national parks to protect this natural ecology with all its variety and change. National parks should not be tamed, subdued, and managed into units as prosaic as tree plantations."
11. Rick Harmon, *Crater Lake National Park: A History* (Corvallis: Oregon State University Press, 2002).
12. Ibid., 187.
13. Ibid., 211.
14. Ibid., 212.
15. Ibid., 263. Also see Doug Larson, Clifford Dahm, and Stan Geiger, "Limnological Response of Crater Lake to Possible Long-Term Sewage Influx,"

in *Crater Lake: An Ecosystem Study*, ed. Ellen T. Drake et al. (Washington, DC: American Association for the Advancement of Science, 1990), 197–212.

16. Douglas Larson, "Neglecting a National Treasure," *Oregonian*, March 17, 2009.
17. Harmon, *Crater Lake National Park*, 216.
18. Edward E. C. Clebsch, "Concerning the Values to Science of Wilderness in the Great Smokey Mountains National Park, North Carolina–Tennessee," June 23, 1969.

CHAPTER 34. THE VAIL CALL TO ARMS, UNHEARD

1. Nathaniel P. Reed, draft of a manuscript, 2009.
2. *Washington Post*, May 15, 1989.
3. *Los Angeles Times*, April 30, 1989.
4. James M. Ridenour, *The National Parks Compromised: Pork Barrel Politics and America's Treasures* (Merriville, IN: ICS Books, 1994), 77–78.
5. Ibid., 79.
6. Ibid., 162.
7. Ibid., 79, 107.
8. James G. Watt, response to a question about environmentalists in an interview; reported in *Forest Industries* 109, no. 4 (1982): 21–23.
9. *Washington Post*, February 11, 1985.
10. *Tribune* (Oakland), March 11, 1985; *San Francisco Chronicle*, February 9, 1985.
11. Mott, interview with the author, Washington, DC, 1986.
12. See Michael S. Lasky, "Can He Restore Our National Parks? Director William Penn Mott Is 76, but He's Going Like 60," *Parade Magazine*, November 24, 1985.
13. According to the *Los Angeles Times* of February 5, 1987, National Park Service Director William Penn Mott gave Western Regional Director Howard Chapman an efficiency rating of "level 2" in fall of 1986. On a scale of 1 to 5 (1 being best), Mott's evaluation indicated that Chapman had "exceeded performance standards." Horn wanted to change Chapman's rating to an "unsatisfactory" level 4, and recommended Chapman be immediately reassigned. On resistance from Mott, Horn compromised on a level 3, or "average," rating. Chapman resigned May 2, 1987. See also *High Country News*, June 22, 1987.
14. *Anchorage Daily News*, June 16, 1988.
15. *Sacramento Bee*, December 9, 1986; *Federal Times* (Washington, DC), January 12, 1987; *High Country News*, June 22, 1987. See also Edward A. O'Neill, *Rape of the American Virgins* (New York: Praeger Publishers, 1972), 92.

CHAPTER 35. "THEIR LABORS WERE NOT IN VAIN"

1. Statement by Rep. Bruce Vento, 1996.

CHAPTER 36. COULD MY WORDS POSSIBLY MAKE ANY DIFFERENCE?

1. Marjory Stoneman Douglas, *The Everglades: River of Grass* (St. Simons Island, GA: Mockingbird Books, 1989), 300.
2. Ibid., 5–6.
3. Nathaniel P. Reed, Testimony on Everglades Restoration before the United States Senate Environment and Public Works Committee Field Hearing, January 7, 2000.
4. William E. Gibson, "Glades restoration money thrills conservationists," *Sun Sentinel*, December 21, 2011.
5. Stoneman Douglas, *The Everglades: River of Grass*, 308.
6. Among public comments the Park Service received on its draft prairie dog management plan for Badlands National Park in 1979 were the following:

> The Pennington County Commissioners are pleased to learn that you are proposing to control prairie dogs. It is our opinion that prairie dogs are nothing more than a rodent (prairie rat might be a more appropriate name) and to continue to allow them to infest good grazing land and to re-infest private land is beyond our comprehension.
>
> I am a producing farmer and rancher in the adjoining neighborhood and this is affecting my livelihood. The Park Service is causing my operation undue loss because they are not concerned for my future or the future of the consuming public for whom I produce food. We have faced the influx for five years with an unending and unsuccessful fight to control the dirty dogs. We are losing the war. We cannot continue to spend thousands of dollars out of our own pockets for a problem that started with the Park Service.

7. Harvey Manning, *Wilderness Alps Conservation and Conflict in Washington's North Cascades* (Bellingham, WA: Northwest Wild Books, 2007), foreword by David R. Brower; edited by Ken Wilcox for the North Cascades Conservation Council.
8. John Ellis, "New plan for Merced River in Yosemite," *Fresno Bee*, October 1, 2009.
9. Eugene Rose, letter to the editor, *Fresno Bee*, April 3, 2008.
10. John Muir, "The Tuolumne Yosemite in Danger," *The Outlook*, vol. 87 (September–December 1907): 486.

CHAPTER 37. WHEN A WHISTLEBLOWER "GOES PUBLIC"

1. Memorandum from Chief of Park Maintenance to Glacier Superintendent, August 12, 1971.
2. Habeck quote from unpublished manuscript for *National Parks* magazine.
3. *McClelland v. Andrus*, 606 F.2d 1278, 1285 (D.C. Cir. 1979).
4. Gary Everhardt, interview with the author, Waynesboro, Virginia, 1985.

CHAPTER 39. HORACE ALBRIGHT TYPING ON HIS AGED PORTABLE

1. Robert Engberg, ed., *John Muir: Summering in the Sierra* (Madison: University of Wisconsin Press, 1984), 121.
2. John Muir, "Hunting Big Redwoods," *Atlantic Monthly*, vol. 88 (July 1901): 311.
3. Two years later, in 1986, twenty-five of the Horace M. Albright Lectures in Conservation were published in a collection by the University of Idaho Press titled *Conservators of Hope*. These lectures—including Albright's "Great American Conservationists" (1961), Barry Commoner's "Ecology and Social Action" (1973), Ansel Adams's "The Role of the Artist in Conservation" (1975), and David Brower's "Conservation and National Security" (1982)—give a review of a period in time, with its problems and prospects.

CHAPTER 40. THE COALITION BRINGS ITS EXPERIENCE TO THE TABLE

1. Teresa Chambers, former chief of the U.S. Park Police, email to supporters, November 2008.
2. An editorial in the *St. Louis Post-Dispatch* of November 26, 2008, doesn't tell it all, but it tells it well, as evidenced by these excerpts:

> As the clock ticks down to Inauguration Day, Jan. 20, the Bush administration is working feverishly to dismantle at least 10 major safeguards of the nation's air, water, endangered species and national parks.
>
> Most of the damage took place before Nov. 15, 60 days before Inauguration Day. That's because most new federal rules take effect 60 days after being published in the *Federal Register*. Once in effect, they are more difficult and time-consuming to undo. In recent weeks, the Bush administration has:
>
> - Opened up 2 million acres of Western land to the development of oil shale, one of the dirtiest fuels on the planet. Another 360,000 acres—including large swaths of public land near Arches National Park and Canyonlands National Park in Utah and Dinosaur National Monument on the border of Utah and Colorado—were opened to oil drilling.
> - Exempted large factory farms and mountaintop mining operations from parts of the Clean Water Act. The prohibition against dumping mining waste into rivers and streams dates back to the administration of President Ronald Reagan, who was not exactly an environmental radical.
> - Loosened clean-air rules to make it easier to build power plants, refineries and chemical plants near national parks. It also changed rules to make it easier for coal-fired power plants to avoid installing pollution controls or clean up soot and smog emissions.

- Changed rules to prevent Congress from blocking uranium mining on claims filed near the Grand Canyon. Higher prices for uranium have prompted hundreds of new mining claims on federal land. In June, a House committee ordered that about 1 million acres of land near the Grand Canyon be exempt from mining. The rule change would block that.

3. Coalition of National Park Service Retirees, *A Call to Action: Saving Our National Park System*, September 21, 2004, 15.

CHAPTER 41. ARTISTS AND PHOTOGRAPHERS DIRECT US TO A SENSE OF PLACE AND OF SPIRIT

1. Devereux Butcher, *Exploring Our National Parks and Monuments*, 5th ed. (Boston: Houghton Mifflin Company, 1956), v.
2. Rockwell Kent, *Wilderness: A Journal of Quiet Adventure*, Foreword by Doug Capra (Middleton, CT: Wesleyan University Press, 1996), xi.
3. Alan Gussow, *A Sense of Place: The Artist and the American Land* (San Francisco: Friends of the Earth, 1972), 133.

CHAPTER 42. BROWER, WITHOUT FEAR OR FAVOR

1. Alex Barnum and Glen Martin, "Sierra Club Legend Dies: Environmentalist Was Uncompromising Steward of the Planet," *San Francisco Chronicle*, November 7, 2000.
2. Gordon Robinson, *The Forest and the Trees: A Guide to Excellent Forestry* (Washington, DC: Island Press, 1988).
3. Brock Evans, "Hells Canyon Story, Part IV" http://www.hostgeni.com/host-info/hellscanyon.org.
4. David Brower, foreword to an earlier version of the book. Harvey Manning, *Wilderness Alps: Conservation and Conflict in Washington's North Cascades* (Bellingham, WA: Northwest Wild Books, 2007), 479.

CHAPTER 43. FAILING TO SAFEGUARD THE SACRED, ANCIENT, AND FRAGILE

1. Stephen Trimble, "The Center of the Spiral," in *Voices from a Sacred Place: In Defense of Petroglyph National Monument*, ed. Verne Huser (Berkeley: Sacred Sites International Foundation, 1998), 53.
2. J. J. Brody, "Petroglyph National Monument Still in Peril," *La Pintura* 30, no. 3 (April 2004): 10.
3. Harvard Ayers, personal interview with the author, 2009.
4. Paul Berkowitz, *The Case of the Indian Trader: Billy Malone and the National Park Service Investigation at Hubbell Trading Post* (Albuquerque: University of New Mexico Press, 2011), 59.

CHAPTER 44. SOMETIMES RULES AND REGULATIONS ARE BENT AND BROKEN

1. Sellars was not the only one to speak in these terms. Jim Walters, wilderness program coordinator of the Intermountain Region, also complained

that the agency was not meeting its responsibility. When he was ordered to transfer, he chose to retire and dispatched this memorandum to the director on January 22, 2004:

> Currently, the NPS administers the nation's, and in fact the world's, largest wilderness inventory. This inventory includes 45 park areas containing designated wilderness and, at least, 31 additional areas which contain lands which have been identified as "recommended," "proposed," "potential" and "suitable" wilderness resources. (This inventory is acknowledged to be incomplete in that many national park areas have never completed even the basic required assessment for potential wilderness designation.) Together these lands represent approximately 86% of National Park Service lands.
>
> The NPS consequently remains vulnerable to growing criticism from the environmental community that: (1) the agency has failed to properly identify and protect its wilderness resources, (2) senior level managers continue to demonstrate either a lack of concern and/or an open hostility to the Service's wilderness responsibilities, and (3) park managers continuously attempt to ignore or circumvent the instructions of the Wilderness Act and NPS wilderness policies in carrying out their other duties.
>
> In Summary: After 40 years, the National Park Service has done relatively little to demonstrate that it has taken its wilderness management responsibilities seriously nor has it implemented a management program which reasonably provides for the day-to-day and long-term preservation of this resource. The lack of evidence that the Service has met even its most basic responsibilities as required by the Wilderness Act, and its own policies, after this amount of time has generated a growing distrust of the agency by the public, and especially within the environmental community. This distrust is exacerbated by the growing number of incidents throughout the Service wherein NPS staff violate the letter and spirit of the Wilderness Act, and NPS wilderness management directives, with little or no consequences.
>
> The Service's current wilderness program as a whole falls far short of what should be expected from the United States National Park Service. The real measure of the Service's success in preserving wilderness is not going to be found in memos and directives, training classes, brochures, annual reports, conferences and meetings, but what is actually taking place on the ground. While I recognize that some slow progress has been made, primarily through the efforts of the National Wilderness Steering Committee and other dedicated individuals, the reality is that unless the NPS Directorate is willing to provide a better system of accountability for the management of wilderness, these types of products will continue to serve as little more that a facade for an inherently

weak program. Continuing at the current level of management will undoubtedly expose the NPS to further litigation and further dilute the Service's fading image as a steward of the nation's natural resources.

CHAPTER 45. GOING HOME

1. Stephen T. Mather and Robert Sterling Yard, *National Parks Portfolio* (Washington, DC: Department of the Interior, 1916), 1.
2. Nancy Newhall, *A Contribution to the Heritage of Every American: The Conservation Activities of John D. Rockefeller, Jr.* (New York: Alfred A. Knopf, 1957), 177.
3. Ronald C. White Jr., *A. Lincoln: A Biography* (New York: Random House, 2009), 501.

Index